SpringerBriefs in Applied Sciences and Technology

Safety Management

Series editors

Eric Marsden, FonCSI, Toulouse, France
Caroline Kamaté, FonCSI, Toulouse, France
François Daniellou, FonCSI, Toulouse, France

The SpringerBriefs in Safety Management present cutting-edge research results on the management of technological risks and decision-making in high-stakes settings.

Decision-making in high-hazard environments is often affected by uncertainty and ambiguity; it is characterized by tradeoffs between multiple, competing objectives. Managers and regulators need conceptual tools to help them develop risk management strategies, establish appropriate compromises and justify their decisions in such ambiguous settings. This series weaves together insights from multiple scientific disciplines that shed light on these problems, including organization studies, psychology, sociology, economics, law and engineering. It explores novel topics related to safety management, anticipating operational challenges in high-hazard industries and the societal concerns associated with these activities.

These publications are by and for academics and practitioners (industry, regulators) in safety management and risk research. Relevant industry sectors include nuclear, offshore oil and gas, chemicals processing, aviation, railways, construction and healthcare Some emphasis is placed on explaining concepts to a non-specialized but still academic audience, and the shorter format ensures a concentrated approach to the topics treated.

More information about this series at http://www.springer.com/series/15119

Corinne Bieder · Claude Gilbert
Benoît Journé · Hervé Laroche
Editors

Beyond Safety Training

Embedding Safety in Professional Skills

 Springer Open

Editors
Corinne Bieder
École nationale de l'aviation civile (ENAC)
Toulouse
France

Claude Gilbert
CNRS
Grenoble
France

Benoît Journé
Université de Nantes
Nantes
France

Hervé Laroche
ESCP
Paris
France

ISSN 2191-530X   ISSN 2191-5318 (electronic)
SpringerBriefs in Applied Sciences and Technology
ISSN 2520-8004   ISSN 2520-8012 (electronic)
Safety Management
ISBN 978-3-319-65526-0   ISBN 978-3-319-65527-7   (eBook)
https://doi.org/10.1007/978-3-319-65527-7

Library of Congress Control Number: 2017949502

Printed on acid-free paper

This Springer imprint is published by Springer Nature
The registered company is Springer International Publishing AG
The registered company address is: Gewerbestrasse 11, 6330 Cham, Switzerland

Preface

This book is the fruit of an original project launched by the Foundation for an Industrial Safety Culture, FonCSI, at the beginning of 2015. It was inspired by a question about "professionalization in the field of industrial safety" put to FonCSI by its industrial partners.

Briefly summarized, this question would be:

"Resources devoted to safety training are becoming more important, however it appears that expectations are not being entirely met, particularly in the industrial sectors that have already achieved a high safety level. Why, despite all the efforts made to provide training, in the broad sense of the term, is there no tangible evidence of actual results in terms of safety? Why do accidents still occur? What are the ways forward?"

An Under-Researched Topic

Despite the two themes being widely studied individually, the links between professionalization/professionalism and safety are relatively unexplored by the academic world. Why was this theme not put on the agenda earlier? Why is it little addressed in the academic literature? The answer is probably related to disciplinary issues, but also to the lens chosen to tackle the problem. Industrial safety is a broad multidisciplinary field, ranging from engineering to social sciences, addressing both human behaviour, organizational issues, regulation and more. Skills and professionalism are the subject of extensive works in the area of educational sciences, occupational sociology and human resource research, the latter being located at the crossroads of the others. However, these works are usually disconnected from safety practices. The question that was put before us actually focused on the interface between man, technology and organization, and was likely to mobilize many disciplines and theoretical currents. In light of this complexity, FonCSI was initially rather challenged. The first issue was of semantic origin. What do we mean by professionalization, professional? It rapidly appeared that the meanings greatly

differ especially between France and the English-speaking world[1]. And then
FonCSI thought of a number of other questions to be answered: Who are we talking
about? The safety/HSE[2] professional or any professional operator? Are we
addressing the field of major accident risks, of occupational risks, or both? Does
professionalization in/of safety make any sense? What part could job professionalization play in ensuring safety? Safety at the organizational level: beyond the
individualistic viewpoint of professionalization?

An Original Research Format

Once the question had been clearly formulated, the next stage was to identify the
way to address it. It will come as no surprise to the reader that the industrial concern
—despite being short and clearly worded—could not be resolved by a simple and
unequivocal response, because of the actors and dimensions to which it relates as
well as for the challenges it represents for the present and future of at-risk industries. No, such an issue required special treatment, an innovative methodology: a
strategic analysis. This was conducted by a group composed of scholars from
different academic disciplines and countries, and practitioners from various
industrial sectors such as oil and gas, energy and transports. The group were also

---

[1]The initial title of the strategic analysis group was "*La professionnalisation en sécurité*". The first
international exchanges quickly highlighted that it was impossible to translate this title by "professionalization in safety". In English, the term 'profession' refers to intellectual occupations (such
as physicians, lawyers, engineers), which:

- are closed, in the sense that, to enter them, one should go through a process of authorisation
  and/or certification, the criteria of which are defined by the profession and usually protected by
  public authorities;
- have gained the monopoly of performing certain activities, through arguing that, given the
  social importance of the latter, they should only be entrusted to highly qualified professionals;
- have professional bodies that control their members' integrity and lobby to maintain the
  profession's social status.

In French, the term '*professionnel*' refers more to the idea of **trade**: an occupation with a
collective history, during which its members, through a process of discussion about their practices,
have built 'rules of the trade' that are passed on from one generation to the other, but also
continuously enriched through an ongoing collective reflection on difficult situations and the best
ways to deal with them.

Therefore, although both meanings exist in both languages, in English, as stated by Nilsson in
his lecture "What is a profession?" in 2007, **'*Professionalization*'** *generally is the social process
by which any* trade *or occupation transforms itself into a true*'profession *of the highest integrity
and competence*') whereas in French, **'professionnalisation'** refers to the process by which a
newcomer enters the professional group and benefits from its historical collective reflection in
order to become 'a good professional' more quickly.

[2]Health, Safety and Environment.

able to benefit from the experience of academic experts in human resources[3]. The different disciplinary angles through which the topic has been addressed are as many entry points with different objectives: management science focuses on industrial performance, sociology on the collective groups within the organization and the political dimension of the question, while ergonomics studies operators' actual activity, to mention a few aspects. The originality of the research design lay in the interaction between all these experts as it encouraged them to compare their ideas and ultimately come up with a strong and innovative overview of the subject. This confrontation of viewpoints provided for a richer and better informed debate during a two-day international seminar organized by FonCSI in November 2015. Furthermore, creating and maintaining a long-lasting discussion led to an actual socialization of the experts within the group. This research process achieved its objectives of driving in-depth reflection and providing concrete ways to collectively go beyond traditional approaches to the delicate issue of the link between education, professionalization, competences and safety.

A Unique Production

This book not only reflects the most precious viewpoints of experts from different disciplines[4] and different countries[5] with experience in various industrial fields at the cutting edge of theories and practices in terms of safety, professionalization and their relationships. It also makes optimal use of the high-level discussions that were conducted, and consolidates the positioning of FonCSI in the field of professional development and safety. It highlights what is currently considered to be at stake in terms of safety training, in the industrial world (industry and other stakeholders such as regulatory authorities), taking into account the system of constraints to which the various stakeholders are subjected. It reports some success stories as well as elements which could explain the observed plateau in terms of outcome. It identifies some levers for development for at-risk industries and outlines a possible research agenda to go further with experimental solutions.

Chapter 1 is the introduction to the book. It questions the links between safety and 'professionalization' according to the following dialectic. 'Ordinary safety', means safety embedded in everyday industrial practices where the more professional one is in one's dedicated duties, the safer one works. Yet 'extraordinary safety', namely safety isolated from other working dimensions, is a matter of exception and safety training requires specific actions from specialized departments

---

[3]Valérie Boussard (professor in sociology of work, Université Paris Nanterre, France) and Sandra Enlart (researcher in educational sciences and CEO of *Entreprise & personnel*, France).

[4]Social sciences, psychology, ergonomics, management, political science, educational sciences, engineering…

[5]Australia, Belgium, France, Italy, Netherlands, Norway, United Kingdom.

and professionals. Claude Gilbert thus elaborated on safety to meet internal objectives or safety to comply with external stakeholders' expectations, more as a justification requirement.

In Chap. 2, Silvia Gherardi addresses safety as an emergent property of a sociotechnical system, a collectively constructed organizational competence incorporated into working practices.

Chapter 3 by Pascal Ughetto highlights the tension between central management, acknowledged as specialists in setting safety policies and middle management, which has great knowledge of real work situations encountered by their teams but whose expertise in that domain does not receive enough recognition. To reduce the gap between rules made for work as thought and rules made for work as done, the author demonstrates the importance of reinforcing the role of middle managers in setting the organizational rules of the teams they manage.

In an unconventional Chap. 4, Hervé Laroche plays the devil's advocate, by the means of a fictional dialogue between an operator and a manager, to critically assess the injunction of professionalism that is defended in this book. The aim of this contribution is to stimulate debate and develop alternatives for managers.

In Chap. 5, Pierre-Arnaud Delattre mainly addresses the differences between France and English-speaking countries along two axes. First he describes the differences in terminology of the word 'professional' and related terms, then he shows that their respective approaches of human and organizational factors in OH&S[6] originate from their own specific history.

In Chap. 6, Rhona Flin highlights that rather than specific safety training, integrating safety thinking by addressing workplace behaviour (non-technical skills) and attitude to risk (chronic unease) in routine work are keys to improving both job performance and skills for safety.

By means of empirical study cases in shipping, railways and space operations, in Chap. 7 Petter Almklov analyses the relationship between representations of work (rules, procedures, models, specifications, plans) and the real and contextualised practice of involved professionals. By showing how compartmentalization of safety can disempower practitioners and by discussing the role of procedures and rules, it offers some propositions about the relationship between professionalization and safety and reliability.

Chapter 8 by Jan Hayes suggests keys for promoting and maintaining the 'safety imagination' of experts in order to take into account lessons learned from accidents and near-misses with regards to future decision-making.

With Chap. 9, Linda Bellamy addresses the subject of professional development by opposing two types of 'professionals'. The former, by doing what is right, manages risk in their activities 'naturally' by using their professional skills and expertise; the latter, by complying to standards and procedures, does what is safe. This refers to two distinct manners of approaching safety, safety as embedded in working practices and normative safety. By means of lessons learned from

---

[6]Occupational Health & Safety.

accidents, but also by insisting on the poor attention given to lessons learned from successful recoveries, the author highlights important issues in terms of safety competence development, particularly in the management of uncertainty.

In Chap 10, based on his extensive expertise in the mining industry, Jonathan Molyneux raises the issue of the importance of operational experience, besides acquiring formal safety qualifications, to improve safety performance in at-risk industries. He highlights the paradox by which the influencing aspect of the work of 'safety professionals' as valued advisors is somehow challenged by the fact that they have to meet the compliance agenda and are therefore sometimes perceived by shop floor staff more as a 'procedure-police' than as coaches. Integration versus differentiation with safety improvement strategies tailored for local specific contexts is also discussed.

In Chap. 11, Benoit Journé highlights some inherent contradictions in professional development in risk industries. Neglecting such contradictions would doom training programs to failure. The chapter suggests that bringing safety practices into discussions appears to be a possible way to enhance professional development as well as safety performances.

In Chap. 12, Corinne Bieder addresses the implicit assumptions conveyed by so-called safety training sessions. She unravels them and the underlying understanding of how safety is ensured, thus allowing for better appreciation of what safety training can achieve and, more importantly, what it cannot. She goes beyond these apparent contradictions to offer ways forward for re-thinking 'safety training' and make it an actual lever for enhancing safety performance.

Chapter 13 by Vincent Boccara presents a training design approach oriented by a holistic real-world works analysis based on several works of research. It is about making people able to deal with real-world work situations, rather than them only knowing and applying exogenous standards. Two main axes of progress are identified and could be developed into guidelines for training people to deal with work situations: participatory methods and transformation of both the trainer and the trainee's activity.

In view of the wide and varied offer of theories and methodologies examining human factors in industrial risk, Paul Chadwick, in Chap. 14, proposes a unified approach with a coherent interdisciplinary conceptual framework for both research and intervention. Unlike 'behavioural safety' programs, rather than limiting analysis to behaviour as the root cause of accidents (identification of 'unsafe behaviours'), this approach seeks to influence the contextual elements that explain these behaviours, the 'behavioural determinants'. The approach consists of depicting the situation by identifying why things go wrong and why they go well, and modifying the physical, technical, social and / or organizational context to reduce the occurrence of 'risky' behaviours.

In Chap. 15, Nicolas Herchin focuses on the issues of professional identity and the power of specialists in support functions. His premise is that giving more power and consideration to people in the field i.e. shop floor teams and middle managers is a first step towards an enhanced (safety) performance. This involves a 'liberation' process by which the classical vision of hierarchal structures is reversed, and the importance of learning and knowledge are acknowledged as key sources of motivation.

The sixteenth and last chapter synthesises the main findings from the book and offers avenues for further research.

FonCSI, Toulouse, France   Caroline Kamaté
                                                                        François Daniellou

Contents

Chapter 1
Safety: A Matter for 'Professionals'?

An Introduction

Claude Gilbert

Abstract This opening chapter questions the links between safety and 'professionalization' according to the following dialectics. 'Ordinary safety', means safety embedded in everyday industrial practices where the more professional one is in one's dedicated duties, the safer one works. Yet 'extraordinary safety', namely safety isolated from other working dimensions, is a matter of exception and safety training requires specific actions from specialized departments and professionals. The author then elaborated on safety to meet internal objectives or safety to comply with external stakeholders' expectations, more as a justification requirement.

Keywords Safety training · Safety performance · Professional · External justification

FonCSI's industrial partners made the following observation: training programmes in the field of industrial safety no longer seem to be yielding the results expected of them, despite the attention and funding they receive.

This initial statement merits further discussion. The field of industrial safety is not strictly delimited and clarifying its boundaries would be quite a task in itself. Similarly, the notion of training can be understood in multiple ways. Furthermore, criteria for assessing the true impact of activities described as 'safety training' are lacking. There are no large-scale audits available that would make it possible to determine the role played by these training activities in maintaining or increasing safety and, conversely, what effects might result from their reconsideration. This, in itself, is a difficulty that probably contributes to the unease of industrial companies. Although the latter are convinced that they have identified a problem (insufficient safety training) and a possible solution (make safety training more 'professional'), they wished to explore the matter further.

C. Gilbert (✉)
CNRS/FonCSI, Toulouse, Paris, France
e-mail: claude.gilbert@msh-alpes.fr

© The Author(s) 2018
C. Bieder et al. (eds.), *Beyond Safety Training*, Safety Management,
https://doi.org/10.1007/978-3-319-65527-7_1

Researchers are well aware that if there are problems in search of solutions there are also solutions in search of problems (Cohen et al. 1972) and this can introduce analysis biases. Consequently, during a series of meetings, the academic and industrial experts mandated to explore the links between professionalization and safety endeavoured to reconsider the questions posed by the industrial companies, by mainly focusing their analysis on three points:

- Where do professionalization and safety training meet?
- Should safety training be incorporated into everyday practices and activities or should it be the subject of specific actions within companies?
- Does safety training primarily meet internal requirements dictated by the specific problems companies encounter? Or external requirements dictated by external entities such as regulating authorities, the public, the media, etc.?

1.1   Professionalization and Safety

The debates held within the group identified two different ways of conceiving the link between professionalization and safety (bearing in mind that neither 'professionalization' nor 'safety' are obvious or univocal notions, as explained in the preface):

- the attention given to safety would seem to be closely linked to the skills and know-how that actors engaged in industrial activities learn through the occupations or duties for which they were initially trained;
- the attention given to safety would seem to result primarily from specific actions and training courses which are distinct from the initial training received.

In the first scenario, no particular actions are required in order for safety to be taken into account, as it forms part of the skill-set of the various categories of agents working in industrial companies. It could be said that safety is 'absorbed' into all training (in the broad sense of the term) delivered within and outside companies, so that agents may effectively and safely participate in industrial activities that are essentially risky. From this perspective, maintaining and increasing safety primarily results from the capacity of these agents to be 'good professionals' when carrying out the duties assigned to them, with reference also to the profession to which they belong. In short, safety is 'part of the job' and there is no need to stress that it is a priority because, in a way, it is self-evident for all professionals. One difficulty, however, stems from the fact that this approach to safety seems to apply more to operators and first line supervisors than to middle managers and senior managers.

In the second scenario, taking into account safety relies, as a priority, on specific actions undertaken by individuals and departments specialised in the field of so-called 'safety'. This refers to the continually expanding area of expertise that developed as safety was emerging as an absolute imperative in contemporary societies which, in France and elsewhere, are now identified as risk societies (Beck 1992).

A market has even developed at the crossroads of academic research and consulting, with an increasingly diversified offering based on various safety models. In this context, increased safety is primarily expected to be achieved by increasing the 'professionalization' of these specialists who influence industrial activities by distinguishing themselves from the agents directly involved in the flow of operations. Professionalization goes hand in hand with the identification of a profession or of a group of specific professions, both internally (specialised departments and services, etc.) and externally (training bodies, experts, consultants, etc.). As has been observed in various ways, this can lead to them becoming cut off from the reality in the field.

Depending on the option chosen, 'professionalization and safety' can thus be understood in very different or even opposing ways. Besides the issues with terminology, this can largely explain the ambiguity that often exists in the way these matters are approached. Even more so because, within companies, greater safety can be sought simultaneously on both these levels with varying degrees of visibility.

1.2 Ordinary Safety or Extraordinary Safety

The group also considered how to include the issue of safety in company activities. There again, two main conceptions emerged:

- one which considers that safety is an 'everyday concern' and thus cannot be dissociated from all of the practices, processes and organizational systems on which a company's activity relies. In this context, maintaining a long-lasting safe state in a high-risk activity seems indissociable from the existence of 'routines' or, in other words, without the integration and implementation, within everyday operations, of a set of rules, procedures, but also experiences and non-formalised know-how [constantly and dynamically correcting mistakes and problem areas—cf. (Amalberti 2001)] that limit the human cost of actions for agents and organizations. In short, routine, which is often described as a potential source of deviations and problems, can be a necessary evil within organizations, an essential ingredient for safety that is more managed than regulated (Daniellou et al. 2010; de Terssac and Mignard 2011).
- and another, which considers that safety (just like risks and crises) is a matter of exception and that it can only be achieved through deliberate and repeated actions, located outside of everyday operations, so as to keep attention focused on it at all levels. Safety awareness and information campaigns are emblematic of this approach, serving as constant reminders to stay alert because vigilance can wane. The implication is that we must remain keenly aware of risks at all times in order to avoid them. In this context, the 'routinisation' of practices and procedures is usually perceived as a danger.

The first approach is the one that most corresponds to the reality of the situation within companies. But, quite paradoxically, it is the least known and the one that is not always the focus of investigations in the academic field. Consequently, despite research in the field of ergonomics, in the sociology of work and the sociology of organizations, only a partial analysis has yet been carried out of the way safety is ordinarily guaranteed in high-risk companies. Similarly, the issue of safety is broached more from the perspective of its 'extraordinary failures' than its 'ordinary successes', and this contributes to diminishing interest in the complex processes through which socio-technical systems are usually maintained in a satisfactory or, at least, an 'adequate' state. This explains in part why ordinary safety is increasingly akin to a black box that fewer and fewer researchers are attempting to open (Gilbert 2016).

The second approach is more in line with common sense and with the way safety actions are spontaneously considered in companies. Indeed, it seems obvious that safety cannot be achieved without specialised agents and departments intervening in multiple ways to maintain a state of vigilance or—and this is the reason for quality assurance measures—without the actual processes being strictly backed by administrative procedures. Whether the actions are deliberate and made particularly visible or whether they stem from bureaucratic obligations, the goal is to present the risks by showing them in a way that 'speaks to the conscience'. In fact, any outside observer attending industrial facilities or oil rigs, or observing how transport activities are carried out, is spontaneously led to consider that, actually, safety efforts are mainly the result of repeating instructions. But, despite the visibility and publicity given to these actions, it isn't always easy to determine what impact they really have on everyday operations.

Although they are very different, these two approaches both ask what the effective drivers of ordinary safety are in high-risk activities (knowing that they vary depending on the sector of activity and the company). More particularly, they lead us to wonder what really underpins safety: practices or processes that are part of routines and refer explicitly or implicitly to various safety models. Orders supported by communication campaigns, training courses, certifications aimed at prompting vigilance, at introducing and maintaining a safety culture that is widely shared? They also lead to questioning ourselves about what could enable us to get to grips with the reality of high-risk activities (problematic in the first approach, given the numerous factors to take into account; seemingly easier in the second approach, but there is no guarantee then that they will enable in-depth action on what constitutes the hidden face of these activities). This is a considerable challenge for researchers but also for the actors, because it essentially involves determining how it is possible to achieve a good grasp of high-risk activities in order to improve their safety. And also how it is possible to show that we have a good grasp of these activities—a recurring problem with which all those implementing safety-related actions and policies are now confronted.

1.3 Safety for Whose Benefit? The Inside or the Outside?

It also emerged that the difficulties encountered in defining safety actions and implementing them in high-risk activities were not only due to the 'approaches' used, which refer back to different conceptions or even different 'philosophies' in this area, but also to the existence of a double bind which carries a strong contradiction. On the one hand, these actions must solve specific realities and problems that are characteristic of a company or a sector of activity. On the other hand, they must meet a set of external expectations which are increasingly numerous and codified in societies that are conscious of collective risks.

- When it comes to safety in companies, the primary aim is effectiveness, irrespective of the means used (comprehensive actions through professionalization; ad hoc actions through training). Indeed, whatever the difficulties, the goal is always to try to ensure that these actions are as compatible as possible with actual situations (with a wide range of methods available to achieve this, and actually this explains the variety of training options available).
- But, at the same time, companies must provide evidence (to regulating authorities, various associations, the media and, more broadly speaking, the public) that they are making safety their priority. Yet the administration of this evidence must fulfil the criteria that prevail in debates about collective risks (and more particularly industrial risks) according to the rules that apply in the public arena (Gilbert and Henry 2012). So essentially, such evidence can be given publicly by highlighting the efforts made to fund safety measures, ensure standards, rules and procedures are followed, develop a safety culture, etc. Thus, even though quality approaches can be considered an 'internal' justification method, they are also largely in place to meet 'external' justification requirements (particularly those stemming from supervisory authorities or the evolution of jurisprudence).

Safety actions thus find themselves caught in a contradictory injunction, because they must meet both internal requirements (in terms of effectiveness) and external requirements (in terms of justification). Rather paradoxically, the consequence is that the most in-depth actions—those that are have the greatest influence on practices and processes and those that take into account the diversity of the factors that effectively guarantee safety—are those that are least likely to be of use as evidence for 'the outside'. Conversely, those that are the most aligned with public views regarding risk management (by highlighting formal aspects, respect for values, a sense of responsibility, ethics, etc.) are the most immediately useful for company communication (in the very broad sense of the term). Indeed, they are the ones that most fulfil the requirements for accountability (Ayache 2008), which large organizations and companies that must justify their actions and policies in the public arena have now accepted as an obligation. This explains the difficulties people within the company can encounter when they must elaborate a safety training policy, as is the case for HR managers. The training offered is indeed based in large part on what 'the outside' expects from companies when it comes to safety. Consequently, meeting the actual needs in the field can prove problematic.

The analysis undertaken within the group therefore led to meeting the demands of industrial companies by considerably shifting the questioning about 'professionalization and safety'. Indeed, it asks all involved to note the fact that specific safety training courses are at odds in many ways.

Firstly, without it being said clearly, these training courses can find themselves in competition with the pursuit of safety as it is effectively carried out by the different occupations, the practices, and process activation (in other words, anything that can be qualified as 'professional'). Insistence on the professionalization of safety, or indeed the professionalization of safety-related occupations, only contributes to masking the discreet, yet broad, implementation of the ordinary safety processes that are part of high-risk activities (but do not necessarily dictate how they are carried out). Thus it is difficult to tackle head-on the link that must be established between initial training, the skills upgrades required by the different occupations, and the training focused on safety. Similarly, the limitations of many professional development courses that aim to train employees in designated 'theoretical' situations without sufficiently preparing them for the range of situations they are likely to encounter in real life or teaching them the knowledge they need to develop a pertinent response are overlooked.

Secondly, and this is linked to the first point, training activities most often lead to thinking about safety from the perspective of the exceptional, the extraordinary, as if they were barely conceivable outside of specific activities, separate from everyday operations and, above all, carried out most deliberately by specialists (whether those recognised as such within the company or external trainers). Once again, the consequence of this is to render the return to reality difficult and to make the views introduced from 'the outside' seem out of touch or indeed ineffective (irrespective of how close the trainers are to the agents involved in the activities, and despite the middle road taken by proponents of the quality approach).

Thirdly, as the training activities are also used to demonstrate the willingness of high-risk companies to make safety an (absolute) priority, this can in fact shift their core purpose away from the reality of the company's activities. The goal then is less about achieving effectiveness in terms of the management of these activities and more about achieving effectiveness in terms of justifying the efforts made by a company or a sector of activity. And this all the more so as it is not always possible to be publicly accountable for the conditions in which the effectiveness was achieved, given the many constraints and obligations facing the actors undertaking the risky activities. Thus it is always difficult to acknowledge that although safety is an important requirement it is one requirement among many for companies conducting high-risk activities, and that in 'real life' the managed safety that takes into account the various constraints and compromises to which all activities are subjected overrides regulated safety.

Based on these observations, what are the avenues that might be followed and explored by colleagues and foreign experts invited to an international seminar[1]?

First, it seems reasonable to recommend that every effort be made to 'return to reality' by aligning safety training courses with safety as it is actually practised in high-risk companies. Indeed, if high-risk situations are to be handled with professionalism, it is important to encourage debate (or even controversy) between different professionals with regards to the situations they encounter, the way they interpret them, the risks they see in them, the solutions that seem pertinent to them, and the feedback received on the implementation of these solutions. Taking stock and discussing the handling (technical, organizational, pedagogical) of categories of high-risk situations must therefore be a permanent part of each occupation's duties. Similarly, the very wide range of practices, of situations and of networks and groups of individuals actually involved in carrying out and managing tasks, often external to the companies themselves, must be taken into account. This seems obvious, but as previously indicated, there are many obstacles to aligning the goals of safety training programmes with safety as it is handled in the field (ensuring actual practices are 'hidden' if they appear to be scarcely or not at all compatible with the image of safety held in the public arena).

Next, while encouraging training courses to be based on real conditions, it seems necessary to favour a pragmatic approach by acknowledging the fact that although the current situation in terms of safety training is far from ideal, it corresponds to a 'state of the world' and a 'state of relations' in our society which it is still difficult to change. Hence, however effective safety training programmes are, and however well aligned they are with industrial realities, they participate in the justification work that companies and high-risk activities must engage in. It is therefore illusory to think that they could, or even that they should, have as their only objective the pursuit of effectiveness based on the analysis of actual practices and work methods in high-risk activity sectors. In fact, they have another social function whose importance can only grow, given the requirements in terms of accountability. Furthermore, we cannot completely forget that, in France, continuing professional development, of which safety training is a part, is an element in the compromise reached between management and unions. This is another type of constraint that must be taken into account, because it has an influence on the training available. Consequently, it should probably be acknowledged that safety-related training also acquires meaning when, regardless of the explicit or implicit safety model to which it refers, it achieves a goal that has become essential: showing the importance given to the matter of safety.

This situation can be a source of dissatisfaction, as it leads both researchers and actors to 'make allowances', or in other words to admit that the difficulties encountered in the field of safety training have deep-rooted causes. Shifting practices and work methods back to reality in order to build better training activities,

---

[1]The two-day international workshop mentioned in the preface, organized by FonCSI in November 2015 and highlight of the project that led to this book (editors' note).

whether by greater incorporation into occupations, work groups, or by specific interventions, represents a real challenge given that our societies have distanced themselves so much from 'reality'. Similarly, it is not easy to give up training activities that are deemed out of touch or even inadequate when these prove to play an essential social role. Nevertheless, avenues for progress do exist if, instead of focusing on these limitations, we consider that it should be possible, dialectically, to work with these different aspects to improve industrial safety. Going 'back to reality' and getting as close as possible to ordinary activities could make it possible to question the pertinence of safety training programmes. Conversely, the elaboration of safety training programmes can be an ideal opportunity to encourage those in charge of ordinary activities to report on their actual practices and the compromises they make between various demands; on how they relate to standards, rules and procedures; on the way they shoulder their responsibilities and conceive their code of ethics. In other words, the gap that has appeared between the reality of practices, which are less and less visible from a social point of view, and also less and less 'viewable', and the image of it that is given via various safety recommendations, could provide the opportunity to question the currently accepted approach to safety. The main challenge probably lies in making it possible once again to discuss—including in public—the conditions under which safety is actually guaranteed in high-risk activities.

References

Amalberti, R. (2001). The paradoxes of almost totally safe transportation systems. *Safety Science, 37*(2–3), 109–121.

Ayache, M. (2008). Le rendu de comptes dans l'entreprise: théories et perceptions. *Annales des Mines- Gérer et comprendre, 1/2008*(91), 16–25. doi:10.3917/geco.091.0016.

Beck, U. (1992). *Risk Society, Towards a New Modernity*. London: Sage Publications.

Cohen, M. D., March, J. G., & Olsen, J. P. (1972). A Garbage Can Model of Organizational Choice. *Administrative Science Quarterly, 17*(1), 1–25.

Daniellou, F., Boissières, I., & Simard, M. (2010). Les facteurs humains et organisationnels de la sécurité industrielle: un état de l'art. (FONCSI, Éd.) *Les Cahiers de la sécurité industrielle*, (2010–02). Retrieved from https://www.foncsi.org/fr/publications/collections/cahiers-securite-industrielle/facteurs-humains-et-organisationnels/view.

de Terssac, G., & Mignard, J. (2011). *Les paradoxes de la sécurité. Le cas d'AZF*. Paris: PUF.

Gilbert, C. (2016). Collective risks and crise: from the extraordinary to the ordinary. In L. Peilin, & L. Roulleau-Berger (Eds.), *Ecological Risks and Disasters-New Experiences in China an Europe*. Routledge.

Gilbert, C., & Henry, E. (2012). Defining social problems: tensions between discreet compromise and publicity. *Revue Française de sociologie, 53*(2012/1).

Chapter 2
A Practice-Based Approach to Safety as an Emergent Competence

Silvia Gherardi

Abstract This chapter proposes to look at safety as a collective knowledgeable doing, i.e. a competency embedded in working practices. Therefore, by adopting a practice-based approach to inquire into how work is actually accomplished, we can study how knowing safe and safer working practices is kept and maintained within situated ways of working and talking about safety. The knowledge object 'safety' is constructed—materially and discursively—by a plurality of professional communities, according to specific scientific disciplines, controlling specific leverages within an organization, and talking different discourses. In a workplace, there are competing discourses: technological, normative, educational, economic, and managerial. Therefore, learning safer working practices is mediated by comparison among the perspectives of the world embraced by the co-participants in the production of safety as an organizational practice. Training and learning based on situated working practices presumes the collective engagement of researchers and participants in reflexivity, which can help to bring to the surface the experience knowledge embedded in practicing and transform it into actionable knowledge to produce practice changes. In fact, the engagement of practitioners, their experience knowledge and their care for what they do may enhance workplace resilience.

Keywords Working situation · Communities of practices · Discourses

S. Gherardi (✉)
University of Trento, Trento, Italy
e-mail: silvia.gherardi@unitn.it

2.1 Introduction

The invitation from FonCSI[1] to reflect on professionalization and safety beyond traditional approaches requires a preliminary explication on how the three terms are understood, before addressing my main reflection on their relationship.

The meaning of safety may be constructed in different ways according to the disciplinary background of the researcher and the approach he or she develops. Thus safety may be thought of and represented as a multifaceted phenomenon that enables a pluralistic way of inquiry. Moreover, the understanding of the field of safety should be considered in historical terms, since it is in itself a socio-cultural product of specific societies. For this reason, we have seen that from the study of risk (in objectivist terms), the field moved on to the culture of safety as an organizational dimension, to reliability and resilience as situated practices. In fact, we may say that the study of safety is part of a reflexive science, since the knowledge produced is going to change the object of study and the changed object calls for a renewed way of studying it.

For approaching safety through the lens of a cultural, organizational and practice-based definition, I offer the following formulation:

> Safety is an emergent competence which is realized in practice, which is socially constructed, innovated and transmitted to new members of the community of practices, and which is embedded in values, norms and social institutions. It is the final outcome of a collective construction process, a 'doing' which involves people, technologies and textual and symbolic forms assembled within a system of social relations. In other words, a 'safe' workplace—a 'safe' organization—results from the constant engineering of diverse elements (for example, skills, materials, relations, communications) which are integral to the working practices of the members of an organization. Safety, then, is knowledge objectified and codified in an expertise and circulating within a web of practices. In order to exist it must be performed in, by and through safety practices, i.e. through discursive and material social accomplishments (Gherardi 2006: 71).

When we look at safety through the practice lens we see that:

1. safety is emergent from the working practices of a community;
2. it is a collective knowledgeable doing;
3. it is embedded in the practices that perform it.

This 'lens' has implications for research since it requires researchers to study safety by studying situated working practices and how practitioners achieve or fail to achieve safe working practices. In other words, safety has to be understood and explained in context and not treated as decontextualized knowledge that may be transferred from one site to another. At the same time this kind of ethnographic, fine-grained understanding of how safety is achieved in situated working practices constitutes a challenge to theorizing safety across different settings. It is important

---

[1]The two-day international workshop mentioned in the preface, organized by FonCSI in November 2015 and highlight of the project that led to this book (editors' note).

to stress that in implementing safety projects we need a local, contextual and detailed knowledge of how a community of practitioners perform more or less safe working practices, since the focus on safety, as actually done while working, raises the importance of situational improvisation, experience and tacit knowledge as sources of resilience (Johansen et al. 2016).

In proposing a practice lens for looking at safety, we are enlarging the traditional way of looking at safety mainly in relation to prevention and control of processes (or products) related to risk in hazardous activities. When we consider safety as 'knowing-in-practice', we are looking at a kind of knowledge that is pervasive and referring to reliability rather than being limited to risk-related contexts. Any activity should in principle be reliable in its outputs and social effects, especially if we consider that risks are pervasive and prone to happen as a consequence of the growing interdependencies of our 'risk society' (Beck 1992).

FonCSI proposes a large definition of the term 'professionalization' to encompass all kinds of learning and training situations, not limited to traditional classroom training or specific safety-related training. We consider that one of the reasons for this call for papers is dissatisfaction with the delivery of traditional safety knowledge and therefore an implicit issue that needs to be addressed is how this may be imagined and delivered differently. Consequently, I propose to look at professionalization distinguishing three lines of inquiry:

- *Strictu sensu* professionalization has to do with the institutionalization of a relatively new professional figure—the safety manager—. Therefore, the institutionalization of a new 'body of knowledge' in the form of a profession raises questions about the learning curriculum of the aspiring safety professional, the institutions best suited to provide and certify this knowledge, the modalities for inducting the new professional into the organizational culture of the employer and moreover about the role and the activities that a safety manager is supposed to perform within a well-defined context.
- Another understanding of professionalization may refer to a distributed professionalization in which each community of practitioners has mastery of the safety knowledge relative to their own working practices and in relationships with other working practices. When I think in terms of distributed professionalization, we have to examine the issue of how to design training for it in a situated and 'customized' way of engaging the practitioners in continually developing new knowledge.
- Finally, if we consider professionalization as an umbrella term or if we wish to contemplate the actual pedagogy and the de-contextualized safety contents that can be transmitted in a routine way, we have to study safety education plans and their productivity.

Due to space constraints, I shall focus only on the second understanding, at the level of the workplace, leaving aside the other interconnections.

A final consideration for clarifying the positioning of my contribution is what kind of safety training is envisaged when the discourse on dissatisfaction with 'traditional' training is commonly addressed. The training that falls under this category may be considered to be inspired by a bureaucratic logic, aiming to answer to norms of accountability rather than efficacy. Moreover, often training is organized and delivered in an 'ad hoc context', usually in a class and with class modalities and often in multiprofessional contexts to unrelated professional groups. Finally, when we look at the implicit pedagogy of similar training we find that the contents of what is depicted as safety are formed by regulations and laws in the implicit understanding that knowing the regulations will produce different (safer) behaviours.

2.2 Safety as a Collective Knowledgeable Doing

Workplace safety is a particular form of 'organizational competence'. In other words, it is a form of emerging competence sustained in working practices by interactions among various collective actors (Gherardi and Nicolini 2000),[2] and various discourses on what constitutes safety.

What we call 'safety' is the result of a set of working practices shaped by a system of symbols and meanings which orient action but which consist of something more. Safety can therefore be viewed as an emerging property of a sociotechnical system, the final result of a collective process of construction, a 'doing' which involves people, technologies and textual and symbolic forms assembled within a system of material relations. This system of relations is made up of heterogeneous components, and it does not display the traditional distinctions between human and non-human elements, cultural or natural aspects, action and constraints. Rather, all these elements are involved in a constant process of generation called the "engineering of heterogeneity" (Law 1992). A 'safe' workplace or a 'safe' organization are the outcome of the quotidian engineering of heterogeneous elements—competences, materials, relations, communications, people—integral to the work practices.

When we consider safety as a social and collective accomplishment, as something that is done with the collaboration of all the practitioners involved in a working practice, then we can say that it has the following characteristics:

- *It is situated in the system of ongoing practices*. It means that 'safety' cannot be separated from its practice and therefore we have to consider safe and safer working practices instead of studying, researching and intervening on safety in abstraction from its work context.
- *It is relational and mediated by artifacts*. Safety knowledge always manifests itself in social activities sustained by symbols, technologies and relations; i.e. action is always 'mediated'. The essential instrument of mediation is language,

---

[2]This section is based on the theoretical framework developed in Gherardi and Nicolini (2000) and readers are invited to consult it for an in-depth analysis.

and the discursive practices in which action and interactions are made accountable to oneself and to the others. Everyday safety is based on the use of discursive and material artifacts which embody not only practical knowledge and experience but also the history and social relations implicit in the mediating artifact. It follows that safety is performed in, by and through social relations, which are relatively stable and have the capacity to deploy a variety of heterogeneous materials in support of working practices.

- *It is always rooted in a context of interaction, and it is acquired through some form of participation in a community of practice.* The idea that safety knowledge is inextricably bound up with action suggests that we should discard the prejudice that practical knowledge is an inferior form of knowledge. Safety knowledge is competence-to-act, and as such it is primarily tacit and taken for granted, as well as being deeply rooted in individual and collective identity. It is tied to particular circumstances, like for example the need to repair breakdowns in the meaning system on which action is based, or the effort to transfer such competence outside its context of origin. Therefore, safety learning does not consist of the appropriation or acquisition of pieces of knowledge, instead it is viewed as the development of situated identities based on participation, within a community of practice (Lave and Wenger 1991; Wenger 1998). A key element for interpreting safety knowledge in organizations thus is the process whereby novices become part of professional 'worlds', become competent in mastering the jargon and the micro-decisions in the system of social practices which regulate participation in situated working practices.

- *It is continually re-produced and negotiated, and hence it is always dynamic and provisional.* The overall picture, therefore, is one in which safety knowledge is no longer conceived as a stable entity that can be situated in individuals or groups, in technologies or rules; it is instead processual knowledge (knowing) emerging from actions and in constant evolution. Safety knowledge is a provisional and performed set of associations among heterogeneous materials; it is therefore the outcome of a 'doing' which uses as its resources for action such diverse materials as people, technologies, textual and symbolic forms assembled within a social context characterized by the presence of multiple collective and individual actors occupying specific power relations. Safety knowledge is sociomaterial and it is the local product of a craft, based on knowledge resources 'disembedded' from their original context and made available through their transformation, legitimization, institutionalization and circulation. However, these resources are then re-embedded in other contexts, in a process which constantly alters both knowledge and the local context of action.

In summing up, we may say that the engineering of heterogeneous elements involves an effort to integrate modes of action proper to several working practices in the organization and sustained by members who, in that they are engaged in different practices and in different communities of practice, deal with safety in different ways. 'Safety knowledge' therefore takes the form of a 'cultural' competence able to influence the style and manner in which meaning and value are attributed to events and to determine the use to which the resources, technologies, artifacts, and knowledge of a group or organization are put. We can say therefore that the knowledge object 'safety' is constructed—materially and discursively—by a plurality of professional communities, according to specific scientific disciplines, controlling specific leverages within an organization, and talking different safety discourses.

When we examine the many safety discourses, co-habiting the very same organization and none of which are hegemonic or possessing a superior 'truth', we can understand better how workplace safety becomes a contested terrain, which is more often like a 'dialogue of the deaf' than an integration of perspectives. The plurality and contemporaneity of safety discourses has consequences for the learning of safety in a constellation of communities of practice. Learning safer working practices is mediated by comparison of the world perspectives embraced by the co-participants in the production of safety as an organizational practice. We shall develop this argument in the following section, since the comparison among perspectives is made possible by the alignment of mental and material elements, within mutually accountable discursive positions (Gherardi and Nicolini 2002). These alignments are provisional and unstable; they produce tensions, discontinuities and incoherence (cacophony) just as much as they produce order and negotiated meanings (consonance).

2.3 The Quotidian Engineering of Heterogeneous Elements, Embedded in a Plurality of Safety Discourses

The term 'discourse' is used to denote a set of texts able to give a (relative) stable form to an object or set of objects, together with the structures and practices involved in their production and circulation. Discourses are forms of strategic arranging that are intentional but do not necessarily have a subject (Law 1994: 21; Foucault 1980: 95). Discourses are therefore themselves relational effects and, as such, they are necessarily contingent, no matter how durable and established they may appear. To every discourse there corresponds an entrenched action-net of alliances which facilitate translation and mobilization of knowledge and modes of knowing. In the case of safety, there are competing discourses: technological, normative, educational, economic, and managerial. The first three will be illustrated in the next sub-sections, while the latter two considered to be implicit in the logic of the chapter.

2.3.1 Safety Within the Technological Discourse

The 'technological' discourse of safety is matched by a network of institutional actors which comprises, amongst others, engineers, physicists, planners, legislators, producers and distributors of organizational learning practices and products.

Though formally independent, these actors operate in close contact with each other, because they have well-established channels of communication and because they sustain common and complementary practices of organizational learning which are not limited by formal organizational boundaries. Acting as a whole, they sustain the technological discourse of safety that is well expressed by the designer of safety devices who explains that it is possible to "build safety into the equipment, the work and the machinery". This safety discourse reveals a specific understanding of the issue and a specific manner of interpreting and explaining events and actions, and working to encourage or prevent them. Think for example of how the capacity of an artifact (or a technology) to exert its control at a distance, depends on the well-established alliance between the discourse of safety—the use of 'safe' artifacts —and the bureaucratic and repressive discourse of safety.

2.3.2 Safety Within the Normative Discourse

The normative discourse is asserted mainly by governmental or para-governmental control and prevention agencies and by the judiciary. Though formally independent, these agencies operate in close contact with each other: together they constitute a crucial node in the circulation of safety knowledge in any industry. They derive some of their importance from the fact that they occupy a central position in the perpetuation of the dominant bureaucratic discourse on safety. The conception of safety asserted by the control agencies is based on the idea that safety results from the correct application of rules and from obedience to regulations.

For these agencies, the promotion of safety hinges on control and on information about the rules. The alliance between the technological and the normative discourses on safety is made manifest in the support that the control and prevention agencies provide for the artifact, in order to reinforce its capacity to exert control at a distance and to alter ongoing practices, and thereby generate 'safety'. The interpretative flexibility of technology thus becomes an arena of conflict in which the premises of action imposed by the artifact and the action net that sustains it are rejected.

A typical first refusal strategy is an attempt—often successful—to adapt the artifact to routine practices, thereby thwarting (and traducing) the intentions of its designers. To forestall such manoeuvers of translation by users, the technology, and with it the entire action net that has brought it into existence, must ally itself with the control and prevention agencies in order to discourage 'interpretation' by alteration. Through the work of inspectors and controllers, the technology 'mobilizes' all the coercive power imparted by the institution of control and prevention, as well as that of the judicial system, to discourage the 'decomposition' of the device and its material reinterpretation in everyday practices. The alliance is institutionalized in 'industrial standards' of shape and use, giving rise to specific intermediaries in the form of statutory rules, inspections, testing processes and certificates that show that an item meets legal standards.

Another way in which the vigilance and prevention agencies back up an artifact's ability to exert control at a distance consists in their efforts to neutralize a further, very elementary but extremely effective, strategy of resistance: simply ignoring the artifact or the intermediary (for example, by not carrying out one of the tests prescribed). This disregard may be deliberate—claiming economic reasons for not purchasing new technologies which 'meet the legal standard'—or non-deliberate and due to simple ignorance. In both cases the shared goal of 'technology safety experts' and the control agencies is to enforce the use of items which in turn produce a 'control' effect.

The deliberate ignoring strategy is usually dealt with by inspections and controls. Such an enforcement strategy however prefigures new alliances and new manoeuvers in the process of engineering heterogeneous elements and communities of practice.

The representatives of the users of the machinery may come together and employ lobbyists who argue that adopting the technology is economically damaging to companies, so that the law must be watered down or postponed. Enforcement therefore is often backed up with other motivation discourses such as that of 'progress' or 'workforce well-being'. The manoeuver constitutes an effort to enrol other actors in the dispute, who will use the issue for their own purpose: the workers' unions to reaffirm their role as defenders of the rights of workers, and entrepreneurs to gain legitimacy as 'modern and progressive'.

In this scenario, other actors also come into play, who have been 'mobilized' to enforce the use of safe equipment. For example, the firms manufacturing the technology are pressed into service. It is obviously in their interest to argue that safety levels should be improved, since this provides them with opportunities to sell a new generation of products, thereby increasing profits. Their commercial representatives thus become the brokers of the normative discourse, which they assert in order to generate sales. Simultaneously, however, they also act unknowingly as the intermediaries of the knowledge and culture embodied in the artifact.

2.3.3 Safety Within the Educational Discourse

The institutionalizing effect of the control agencies and their system of mobilizations and alliances frequently leads to the involvement of agencies that sustain the discourse of safety as education and training. Information about the importance of the correct use of the artifact is conveyed by training and retraining courses and is included in manuals and information material.

Inclusion of the innovation in manuals signals the success of previous efforts, but it also exerts powerful influence on its own account. It affects, in fact, a further important actor, namely the novices who, preconditioned during their training, perform micro-translation processes in the workplace. If novices are asked to use sub-standard equipment, they may refuse, enlisting the use of innovation in their effort to construct a work identity which differentiates them from the 'old workers'.

To highlight their difference, they may therefore flaunt the use of innovation, and in doing so, unwittingly act as a further link in a chain of alliances and mobilizations.

2.3.4 Safety as the Effect of Competing Discourses

Therefore, safety can be conceived as the effect of an action net, in which competing discourses coexist: the technological discourse with other discourses, such as that of safety as rules and punishment, of safety as education and training, of safety as profit or loss, and of safety as management and planning. Discourses among specific practices are not directly aimed at reaching understanding and/or the production of collective action, but rather at knowledge mediated by comparison among the perspectives of all the co-participants in a practice. Comparing different perspectives does not necessarily involve the merging of diversity into some sort of synthesis—harmonizing individual voices and instruments into a symphony (or a canon)—but rather the contemplation of harmonies and dissonances may coexist within the same performance.

2.4 Implications for Experimenting in Training

The principles on which to base a pedagogy for training that acknowledges the situatedness of safety knowledge are simple and are consequential to the practice-based approach outlined. In the first instance the object of training and learning has been moved to safe working practices and the recipients of such training become the community of practice that collectively reflect on their working practices and the knowledge embedded in them in order to change or improve their reliability. An implication of such a principle is that training cannot be delivered in a separate time and place, but should consider the workplace as a learning place and address the community of practice dwelling in it. In my experience, the representation of working practices (through video, feed-back restitution etc.) to practitioners may be a useful means to reflect on and change practices.

In the second instance the multimodality around safe working practices has to be acknowledged in order to improve the interpretative flexibility and mutual accountability in practicing and dealing with practical responsibility. One way of understanding the different discourses on safety may be translated as the capacity of participating with competence in a conversation that is characterized by tensions and sometime difficult trade-offs. In other words, the knowledge object 'safety' should be learnt during training as an object of concern and not an object of fact. The difference between a matter of fact and a matter of concern (Latour 2004) is that instead of 'being there', whether one likes it or not, matters of concern have to be liked, appreciated, tasted, put to the test. Matters of concern are disputable, they move, they carry one away, they *matter*. Too often safety is approached in a rational way, and persons are conceived as a non-trustable 'human factor'. On the contrary the simple evidence that persons are concerned by safety and that safety concern persons and society could become the basis for action-learning programs (Eikeland 2012; Eikeland and Nicolini 2011) inspired by care in working practices. Since care cannot be prescribed, nor encoded in some sort of evidence-based manual, the possibility of recognizing what is commonly understood to be care in a work setting, and how an implicit understanding and negotiation of care takes place on a daily basis, may become a starting point for the development of a situated repertoire of caring practices in a workplace. In fact, the idea of what is care (and how people are engaged in 'doing' safety) is silently incorporated in working practices. Therefore, for the development of a situated training program in the workplace it should become an explicit topic for discussion and for collective learning. Safety does not speak for itself, often it is 'done' but not 'seen'.

Practice-based studies have experimented with several methodologies—ethnography, reflexivity, narrativity—for enhancing the formative and transformative role of knowledge embedded in working practices (Boud et al. 2006; Fenwick 2003; Hager et al. 2012; Raelin 2001; Scaratti et al. 2009). The necessary condition for this is the collaboration with practitioners working within the organization and "the challenge is thus to devise new ways of making (and considering) people as the authors of their work. The expectation is that this will enable people to shoulder and contribute to the goals of the organizations they belong to" (Gorli et al. 2015). In fact, the collective engagement of researchers and participants in reflexivity (Cunliffe 2003) can help to bring to the surface the knowledge embedded in practicing and transform it into actionable knowledge (Argyris and Schon 1978). Actionable knowledge—for changing practices—emerges when all actors agree to question the issues that are often taken for granted and are ready to address the contradictions and conflicts that might emerge in the process.

References

Argyris, C., & Schön, D.A. (1978). Organizational learning: a theory of action perspective. *Journal of Applied Behavioral Science*, 15, 542–548.

Beck, U. (1992). *Risk Society. Towards a New Modernity*. London: Sage.

Boud, D., Cressey, P. & Docherty, P. (2006). *Productive reflection at work: learning for changing organizations*. London: Routledge.

Cunliffe, A.L., (2003). Reflexive inquiry in organizational research: questions and possibilities. *Human Relations*, 56 (8), 983–1003.

Eikeland, O. (2012). Action research and organisational learning: a Norwegian approach to doing action research in complex organisations. *Educational Action Research*, 20(2), 267–290.

Eikeland, O. & Nicolini, D. (2011). Turning practically: broadening the horizon. *Journal of Organizational Change Management*, 24(2), 164–174.

Fenwick, TJ (2003). Emancipatory potential of action learning: A critical analysis. *Journal of Organizational Change Management*, 16(6), 619–632.

Foucault, M. (1980). *Power/knowledge: selected interviews and other writings 1972– 1977*. New York: Pantheon.

Gherardi, S. (2006). *Organizational knowledge: The texture of workplace learning*. Oxford: Blackwell.

Gherardi, S., & Nicolini, D. (2000). To transfer is to transform: the circulation of safety knowledge. *Organization, 7*(2), 329–348.

Gherardi, S., & Nicolini, D. (2002). Learning in a constellation of interconnected practices: canon or dissonance? *Journal of Management Studies, 39*(4), 419–436.

Gorli, M., Nicolini, D., & Scaratti, G. (2015). Reflexivity in practice: Tools and conditions for developing organizational authorship. *Human Relations, 68*(8): 1347–1375.

Hager, P., Lee, A. & Reich, A. (Eds.). (2012). *Practice, learning and change: practice-theory perspectives on professional learning*. New York: Springer International.

Johansen, J.P., Almklov, P.G. & Mohammad, A.B. (2016). "What can possibly go wrong? Anticipatory work in space operations". *Cognition, Technology and Work, 18*(2), 333–350.

Latour, B. (2004). Why has critique run out of steam? From matters of fact to matters of concern. *Critical Inquiry, 30*, 225–248.

Lave, J. & Wenger, E. (1991). *Situated learning: legitimate peripheral participation*. New York: Cambridge University Press.

Law, J. (1992). Notes on the theory of the actor-network: ordering, strategy, and heterogeneity. *System Practice, 5*(4), 379–393.

Law, J. (1994). *Organizing Modernity*. Oxford: Blackwell.

Raelin, J. A. (2001). Public reflection as the basis of learning. *Management learning, 32*(1), 11–30.

Scaratti, G., Gorli, M., & Ripamonti, S. (2009). The power of professionally situated practice analysis in redesigning organisations: a psychosociological approach. *Journal of Workplace Learning, 21*(7), 538–554.

Wenger, E. (1998). *Communities of practice: Learning, meaning and identity*. Cambridge: Cambridge University Press.

Chapter 3
Line Managers as Work Professionals in the Era of Workplace Health Professionalization

Pascal Ughetto

Abstract Constructing rules for work that foster both health and safety and efficient production entails, in many organisations, the introduction of procedures, tools and techniques implemented by specialists. The purpose of this is to combat the amateur practices and lack of expertise supposedly found not only amongst employees, but also in their managers. This chapter argues that, on the contrary, field managers possess knowledge about working conditions and are actors who are necessarily involved in organising those conditions as well as the work of their teams. In so doing, they protect employees from or expose them to the real and varying circumstances of work. This is the role that needs to be reinforced in order for safety rules to become a real part of work cultures and working practices. However, the forms of power in organisations increasingly limit the recognition of this expertise in the work of managers. The chapter advocates the importance of giving managers power to set situated organisational rules, instead of making these the exclusive prerogative of central management departments.

Keywords Middle management · Organisational rules · Power

3.1 Introduction

In the last 15 years or so, the increasing salience of issues of workplace health, safety and working conditions has led to systematic efforts to tackle these questions within companies. This has prompted the development of management processes—procedures, tools, techniques—and the use of specialists claiming expertise in these fields, i.e. both an understanding (even scientific knowledge) of these subjects and the mastery of the tools associated with them: professionals in industrial risk, in the prevention of psychosocial problems, etc. The article examines this professional construction of the domain of workplace health and safety and working conditions:

P. Ughetto (✉)
Université Paris-Est, Paris, France
e-mail: pascal.ughetto@u-pem.fr

© The Author(s) 2018
C. Bieder et al. (eds.), *Beyond Safety Training*, Safety Management,
https://doi.org/10.1007/978-3-319-65527-7_3

it argues that the professional space of the specialists is in tension, or even in competition, with that of middle managers or field managers, who do not necessarily enjoy the same recognition in these matters as the specialists. However, it can be argued that the role of these middle managers is crucial, or even that they should be the cornerstone of corporate workplace health and safety policies. It is important to promote their role in organising the work of the staff they manage.

Based on field studies that we conducted on workplace health and safety in a variety of sectors (supermarkets, public housing bodies, hospitals, French Ministry of Finance, etc.), we have developed the following argument: as advanced by the other authors in this book, in order to promote working conditions that are safe and protect the physical and mental health of workers, it is crucial to create professional cultures in which the rules of work are constantly updated and pertinent to the realities and variability of the practical situations encountered. In this, managers play a key role: accountable for the organisational rules and required to achieve targets for production and economic efficiency, the challenge they face is to maintain a balance between these targets and rules on the one hand, and professional cultures on the other, in order to consolidate organisational rules that are also health and safety rules. What is crucial is the way they interpret their role or are encouraged to interpret it: as agents for the local "implementation" of centrally-decided rules, or conversely as autonomous actors with the capacity to adjust these central rules to the constraints and challenges of practical situations encountered by staff, and therefore to professional cultures? Workplace health and safety therefore critically reflect the choices companies make and the power relations within them. In companies where the power of the "official" specialists dominates, central rules are very likely to be forced on field managers with no leeway to make them significant in actual working cultures. The result is to challenge the legitimacy of these managers as pivotal agents of work policies. Companies may be tempted by this approach, which has immediate clarity in terms of the assignment of responsibilities and accountability, and in short-term economic terms. Another approach, in appearance more costly, but actually more productive, is one that asserts the powers of field managers to manage the organisation of work in their "perimeter" and, to this end in particular, to encourage conversations and discussions within communities of practice and the adjustment of organisational rules.

3.2   Professionalizing Workplace Health and Safety?

Awareness of work and working conditions has made a comeback in the last fifteen years or so. This has reinforced a trend towards the construction of work as a public problem and has boosted the production of legal regulations, collective bargaining on these issues and action by companies to prevent work-related risks. The conception and implementation of these actions has encouraged the emergence of specialists who, individually and collectively, have built their career and research around what they believe to be an accurate and serious recognition of the problems and expertise that these entail. This has coincided with the development of specialised management processes focusing on the significant technical and legal facets of these matters, and the use of specialists (in psychosocial risks, industrial risks, etc.); in other words, with the professionalization of workplace health and safety, in the sense that this issue is seen as one that should no longer be left to amateurs, but entrusted to people with expert knowledge, specialists in the field. However, there is nothing immediate, technical or neutral about the tackling of work-related issues. This activity is a social construct, equally reflecting the dynamics of the way these questions are constructed as relating to problems that the actors or a sufficient number of them agree to recognise as real, institutional dynamics that encourage the management of these issues and guide the ways they are tackled, and the dynamics of the production of responses (legal texts, negotiated agreements, instruments…). These responses themselves reflect professional dynamics: how professional groups address these questions and participate in the formulation of the problems and the development of the solutions, or indeed claim to be those most capable of implementing those solutions.

For all these reasons, the term professionalization seems applicable, but its use is not straightforward. Its meaning varies considerably from one country to another. The social construction of what are called instituted professions, in the strongest sense of the term, follows different trajectories, with the result that the same activities are not always fully recognised as the attributes of a profession (Wilensky 1964; Neal and Morgan 2000). However, the question here is whether all the actions required to foster health and safety at work must necessarily entail specialist knowledge and the techniques specialists may propose, as experts external to the work activities concerned, or whether instead, one should resist granting excessive autonomy to this knowledge and these techniques. With regard to the position argued in this chapter, the more concrete issue is the potential tension between specialists who claim to be official professionals in this domain and the middle managers who could assert their own standing as professionals with regards to the working challenges of their staff.

If the term professionalization has its uses in this regard, it is because it can cover a multiplicity of meanings, even in a given country and language. It can have at least three definitions, all of which show that this professional status is less a matter of fact than the result of efforts to obtain recognition for a distinct status: successful for some groups, unsuccessful for others, even though they all seek for such a recognition. Professionalization can thus be defined as:

- establishment of a professional group, with its own territory (a jurisdiction), access to which is confined to its members. These members possess prerogatives based on peer-validated expertise; the specialists we are concerned with here probably do not possess all the attributes of a profession, but lay claim to scientific ethics and knowledge in order to persuade others to recognise such a status;

- emergence of specialist professions, proliferation of individuals and of structures equipped with rules to apply in order to do things efficiently, economically, without risks, and therefore differently from amateurs (who are suspected of doing things uneconomically, inadequately or riskily); what is important here is the contrast with amateur action, which is precisely what the specialists who concern us here seek to stress;
- familiarity with an activity, day-to-day practice of an activity, whereby it is claimed as a profession within which the self-reflective practitioner is constituted as an expert. This is the idea that, even for the most apparently simple activities, in a job or in another sphere of day-to-day life, being a novice is characterised by the difficulty of coping with situations that are at first sight quite uncomplicated and that dealing with the activity on a daily basis leads to the development of a capacity which is simultaneously a skilled practice and a knowledge of situations (Gordon et al. 1999). This is quite close to what ethnomethodology reveals about the challenge and the difficulty of dealing competently with ordinary situations.

In the transition from one of these definitions to another, the term ceases to be restricted to groups that have succeeded in having an exceptional status recognised by others, and it becomes apparent that there are other groups, which have not made this kind of social effort or have failed to complete it, whose members nevertheless believe, deep down, that they are in reality specialists and deserve to be recognised and listened to as such. What Jan Hayes writes about professionals:

> Whilst the term 'professional' and 'professions' are implicated in a fairly tangled and unruly web of usage, the characteristics of professionals include being bound by a code of ethical conduct in addition to technical and/or commercial standards and being able to exercise experienced judgement in specific cases, rather than relying completely on application of general rules (Hayes 2014)

is a claim that can reasonably be made by safety specialists; but others (workers, technicians, managers, who are the target of their instructions) would also, no doubt, wish to emphasise that they do not work without ethical standards, technical precision or the exercise of experienced judgement.

All this should undoubtedly be seen as the discursive and practical strategy of a relatively homogeneous group possessing a degree of unity in its representation of reality and its activities. The issue is less about dividing reality between professions that are "really" definable as such, and others that are not, than about understanding the efforts made by groups that are constituted to varying degrees to achieve recognition for the value of their professional activity and their contributions, efforts in which some groups are more successful than others. From the perspective of symbolic interactionism, it is not possible to decide definitively and objectively which groups should be recognised as professions and which should not. Instead, there are professional dimensions that run through all the groups, but their specific nature as a group varies and these forms of professionalization also differ in their robustness. There are struggles, both individual and collective, of varying intensity, to achieve recognition for the activity practised, for its value to the common good and for its value as a complex and expert activity, and to be awarded a status which differs from that of other activities and professional groups.

From this perspective, there exists a whole continuum and a possible link with the fact that even individuals unfavourably placed in the division of labour—at the bottom of the symbolic ladder of working roles—can, in day-to-day conversation, claim the status of experts more capable than others of deciding what needs to be done in their activity. Cleaners who dispute the relevance of instructions about the order in which they should carry out office cleaning tasks or about the detergents they should use, may speak in a way that reveals that they see themselves as professionals in the work they do, because they deal with it every day.

Professional rhetoric generally consists of a back-and-forth, and a connection, between advocating an idea of the general interest, the common benefit of the enterprise, which the group perceives itself as representing, and the development of techniques, methods and tools, in which the group sees the mark of a managed activity. As emerges in the second definition, professionalization is thus seen as taking the place of potentially damaging amateur—or even "cowboy"—practices. Professionals see themselves as having the correct understanding of the issues and tested and safe practices for tackling them. Groups that are close to being recognised as professionals obviously press the advantage to the point of claiming authority to legislate on the practices of those whose place within the symbolic division of labour makes them less able to assert this status.

In other words, the dynamics of professionalization are processes in which there is dispute over the terrain of expertise and over the legitimacy to rule on what should be done and how it should be done, at several levels: moral constructions of the world, the efforts of a group to extend the scope of its prerogatives and autonomy. All this leads to a confrontation of points of view, arbitrated by authorities—company management and legal authorities.

In short, what is at play in professionalization is the rivalry between social worlds to define a negotiated order and establish their place in it. The whole process is in dynamic tension and is never entirely frozen: the groups that lack the power to persuade others to recognise them as possessing genuine professional expertise, will inevitably take advantage of circumstances to show that situations which could have turned out badly were rescued, in fact, by their expertise; specialists, ever threatened by the possibility that the relevance of their knowledge will be contested in concrete conditions, will also seek to exploit opportunities to consolidate their legitimacy; middle managers will try to get their position recognised on occasions where the full value of their experience can make itself felt.

3.3 Specialists Versus Middle Managers

Working safely or adopting rules that ensure health and safety in the workplace is not an objective reality, but more about the conflict between social worlds over the relevance of those rules. Take the case of butchers working in a supermarket, who are subject to a compulsory rule to wear a metal mesh safety glove whenever they cut meat. In any observation of the actual work, it would quickly become clear that there are situations where the butchers fail to wear the glove, to the great displeasure of the safety specialists, who would remind them that there are no exceptions to this rule and no excuses will be accepted. They are very likely to perceive this as the thin end of the wedge, the start of a slippery slope. So, the specialists would see themselves as obliged to turn a deaf ear to the professional arguments, which they see as potentially fallacious. Indeed, the stronger the recriminations, the more convinced they would be that they are right and duty bound not to relent. The butchers, for their part, would be equally sure of being in the right, because of their day-to-day familiarity with the tasks to be done, their experience of the realities and the knowledge they have acquired about the different situations, for example which tasks are easy or difficult, how to handle problems. They would claim that, when you do the job, you know that there are cases where wearing a glove is inefficient and is not necessary because, let us say, there is no risk of injury if the job is done properly.

Working leads to the development of a strong sense that one is ultimately best placed to know how things are and should be. Among workers, the activity gives rise to a professionalization that imparts the feeling that the individual and the group know better than anyone else how people should act and take precautions: this includes the development of know-how not only about self-protection but also about risk.[1] Conversely, the specialist believes that it is his professional duty not to give way to this rhetoric… even if it means denying other people recognition of their professional skill. In the first half of the 20th Century, work specialists, drawing on the different developments in a science of labour (industrial hygiene and psychotechnical methods, ergonomics…), claimed to act to the benefit of the health and safety of workers, even when opposing those workers' typical working practices. Scientific fields and corporate practices gave rise to a debate between different theories of work (e.g. as a biomechanical operation or an activity). Since the interwar years, a trend towards professionalization has arisen, i.e. the emergence of new professions (psychotechnical…) and towards the construction of knowledge and techniques, in opposition to the knowledge of workers but also of team foremen. In the last 20 years, the movement has intensified, reflecting the themes of industrial and psychosocial risk. With the comeback of issues relating to work, working conditions and workplace health and safety—very salient in the French situation since the 2000s—we seem to be seeing the establishment of a specialised process for managing workplace health and safety within the human resource management function.

Between workers and these specialists stand the middle managers. In the case of the supermarket butchers, this would be a section manager. Both of these social worlds "naturally" expect that the field manager will be on their side: the official

---

[1]See, for example, in the French language literature, the collective defence strategies identified by Dejours (1980) or the know-how of prudence analysed by Cru (2014).

specialists will remind him that it is his job to embody the organisational rules, which take precedence because they have been set by the employer and are carefully considered; the work teams will tell him that he is best placed to see that abstract rules decided by specialists who are unfamiliar with the practical local conditions of operation are not always able to ensure compatibility between the combined demands for both productive efficiency and compliance with safety conditions.

Middle managers are often disparaged in current workplace health and safety policies. Whether they are asked to contribute to policies for reducing stress, diminishing musculoskeletal problems or combating workplace accidents, management and human resource departments see them as being in the frontline. If an accident occurs, they are likely to be blamed for failing to supervise the practices of their teams or to place sufficient stress on safety instructions. The successors of the interwar foreman—operations managers, field managers—are often accused of failing to "notice in time", of lacking expertise in "spotting" (e.g. psychologically fragile employees), or of having poor man-management skills. They are therefore required to undergo training (e.g. in spotting fragile personalities) and to comply with centrally set rules for health and safety management processes. The workplace health and safety policies developed by human resource departments often include central training sessions, seen as the solution for raising awareness of the importance of these matters among middle managers and of ensuring that they apply the related organisational rules to the letter. The aim is to make them reliable agents of the effort to organise work both efficiently and safely. In reality, the way in which companies set their work-related policies is based on implicit theories of work: not only the work done by basic employees, but also by the people who manage them. The role of these managers is conceived as being to relay standards as faithfully as possible. Field managers are seen as the frontline representatives of the organisational systems, with the role of implementing those systems, ensuring that they work properly and improvising final adjustments to guarantee that everything runs smoothly. They are expected to act through meetings with their teams, where they relay the right messages and insist on compliance with procedures.

In this respect, they are not identified as specialists in the work of their staff. Work-related expertise is assumed to be central expertise, of which they are simply the vehicles. The training sessions are precisely the times when, it is supposed, they can assimilate that expertise. Whereas they make daily decisions about the work of their teams and, in so doing, become familiar with that work and with what their staff need to carry out their tasks, they are denied possession of legitimate expertise about the work. Their familiarity with the work is sometimes blamed for reinforcing the tendency of staff to "resist change". To what extent central departments and their functional management tolerate field managers organising (and not simply implementing central organisational plans at local level) therefore lies at the centre of the (disputed) social construction of work policies. What boards and central management departments do with the fact that these middle managers know a lot about the work—about the way day-to-day production challenges are handled—

because they manage and organise, dictates what role is allocated to field managers in the rival processes of professionalization in the sphere of work policies.

Safety is thus defined in the interplay between territories and powers through which these three groups of actors—basic workers, their line managers and the representatives of the support functions—seek to achieve mutual recognition. By deploying the formal and informal organisational resources that they are able to establish for themselves, each group tries to use its influence to ensure that its construction of the world and its practices gain ascendancy.

3.4 Middle Management and Functional Departments: The Contested Terrain of the Power to Organise

This prompts us to introduce a new actor into the analysis: the functional departments, for example human resource departments, to which some of the specialists concerned may be attached. Big integrated companies consist not only of a line management structure, but have also developed a management apparatus that includes support functions. Support functions emerged from the division of corporations into specialist functions, but also from the entry into these areas of professional communities keen to have their specific expertise and necessity recognised. The role of functional departments is to develop the standards and tools that enable such a company to survive as an integrated company, i.e. to harmonise operations and guarantee results. Ultimately, they feel responsible for the fact that the company is properly organised. In this context, their representatives expect the field managers to be the vehicles of this organisation, to "implement" it meticulously.

The support functions do not necessarily have hierarchical power over the field managers. Their power is exercised through the setting of standards, for the purpose of harmonising operations, spreading "good practices", ensuring compliance with legislation and maintaining compatibility between decentralised actions. In the sphere of workplace health and safety, the standards are supposed to create the conditions for productive performance and worker safety. They are designed to organise by providing the best ways for the two objectives to coexist. Field managers are then expected to apply them within their own perimeter of authority.

However, that is not all that the field managers do: they are not passive intermediaries who implement organisational rules. They juggle between organisational principles and rules, on the one hand, and field realities within their sector, on the other. They make adjustments, in the knowledge that organisational rules have to be interpreted and adapted to the real activities of their staff. They accept or reject accommodations with the rules. In short, they end up performing an organising role. In this capacity, they do more than to apply to their own teams the organisational frameworks developed by the support functions; in turn, they also contribute to creating these frameworks and to organising the work of their staff. Through these frameworks, staff are protected or exposed (to dangerous machines, to assaults from the public, etc.), the work is made easier or more difficult.

In this face-off between two claims to organise (by the support functions and by the field managers), the tendency in the last 25 years has been for the functional departments to gain increasing power and to gain the ascendancy, for example through technical systems such as IT tools, which impose their underlying formats and rationales.

However, what is in play is the theory of work espoused by health and safety practitioners and field managers: is the aim to comply as closely as possible with requirements that are supposed to guarantee efficiency of production and worker safety; or to develop an activity that needs to be organised, an organisation implemented both by centrally defined rules and instruments and by local adjustments? In the former case, professionalizing health and safety means building up a group of central experts, who will develop tools that operatives and their managers must faithfully implement. In the latter case, professionalizing health and safety means increasing the capacity of field managers to construct organisational solutions that incorporate health and safety preoccupations and know-how of two types: those developed by specialists and those worked out rapidly on a daily basis according to the job and its particular characteristics.

3.5 Concluding Remarks

Companies spend a lot of money on safety processes, but the results of those processes are limited. Most of the chapters in this book argue for it to be recognised that, even when perfectly planned, tasks will always be carried out in contexts that require individuals and groups to improvise to varying degrees; workplace health and safety rules cannot be totally fixed in advance without drawing, at least partly, on the knowledge that individuals and groups develop and exchange about the real activity, its risks, its opportunities, and therefore without adapting to actual working cultures. The aim is certainly to improve processes that enable collective structures to continually develop working rules that are simultaneously rules of efficiency and of safety, and are relevant to the situations actually encountered. It is not to argue that working does not require procedures or that only bottom-up procedures are valid: improvisation is also possible within procedures (Johansen et al. 2016; Almklov, in this volume; Gauthereau and Hollnagel 2005). However, it is important that procedures and standards have instrumental value in the situations encountered. As long as they continue to be perceived as foreign to the realities of work, they are not seen as tools and are therefore not spontaneously applied.

It needs to be recognised that the rules set by communities of specialists will not automatically acquire instrumental value for those required to follow them in their work, and that it is not enough to ask the field managers to transmit and rehearse them. As Silvia Gherardi writes in this volume, a safety culture is acquired as part of a community of practice: this is an integral part of sharing professional identity.

However, it should also be noted that safety rules will not always develop spontaneously in professional cultures, or in the management practices of executives. Referring this time to the suggestion put forward by Rhona Flin, managers may well respond automatically that a good manager will always care about safety, as if there were no need for autonomous thinking or formal conclusions on this specific subject. However, specialised reflection and instruments are not in themselves enough, they need to incorporate communities of practice. Silvia Gherardi, once again, clearly demonstrates that safety rules cannot be learned outside the process of learning to work well. For example, people learn to pay attention to noises as a matter of both skill and self-protection. In Linda Bellamy's contribution, we find the major idea that all this is not purely technical: as a professional doing a job, one first seeks to assimilate the right thing, what one can tell oneself and others is good work, valuable work, even and especially with respect to activities that carry little value in the symbolic division of labour.

What this chapter adds to these ideas is the fact that all this takes place within the power relations specific to today's enterprises. The terrain on which these ideas are propagated is not indifferent to the way in which they reinforce situations or actors, and ask others to evolve in their practices or their power. This does not mean diminishing the status of certain actors to the benefit of others, but changing the terms of the compromises. In power relations within large organisations today, the power exercised by support functions—through the primacy of standards—deprives field managers of a great deal of leeway for action. This power, despite the diversion via participatory management, leaves little room for regular discussion of the relevance of organisational rules. However, rules—in particular safety rules— are not purely and simply "implemented": they need to be discussed (what relevance, what correspondence with actual situations and the real problems that those situations occasion; what effectiveness, what degree of validity, what connection with other professional practices?).

What this work argues for would therefore entail two-way information flows: information that not only flows downwards, but also upward flows of information that are rarely incorporated into managerial decision-making, often more taking the form of complaints from field personnel and their managers, criticisms of existing rules and plans, etc. Central management departments need to be able to hear negatives. They need to be able to interpret opposition and debate as something other than resistance to change. This is anything but simple for central management, which feels challenged in its interpretation of the problems and in its construction of solutions, as well as in its authority, when there is resistance at field level.

This would mean companies that tolerate being what they are: spaces in which there is a multiplicity of points of view, between which there is no immediate accord or even a possibility of accord. What can be done for this to be tolerable, given that a company cannot permit disorder? To achieve this, it needs to be accepted that organising, introducing organisation into day-to-day operations, and notably the organisation of safety, is not about implementing organisational rules and letting them operate unchanged for a given time; the issue is organising, a continual activity of organisation. And this is not a task for the departments officially dedicated to organisation, with field managers simply required to make minute adjustments, but a continuous process of rule creation very largely taking place among these field managers.

To the positions argued by most of the contributors to this book, this article therefore adds that, in following the recommended paths, companies are faced with a choice, which will determine whether they continue along that route or decide to turn back: a choice about the latitude they should give to their middle managers. They need to develop the capacity of these managers to do something with the complaints of their teams, to analyse the work, its constraints, how the teams go about getting things done, and to make proposals to their line managers and their teams. The key question is therefore how much space today's organisations allow for experiment, for variability, and how much space they give middle managers to construct organisational rules, first of all by holding discussions within their teams and between those teams and support departments. As long as companies lack confidence in the capacity of their managers to conduct debate without creating disorder, there will be no alignment with the positions argued in this volume. Today's big organisations also need to combat their fear that debate is an unproductive waste of time. Governing through centralised rules, through standards, together with communication and training to disseminate them, is an apparently more economical solution than having constantly to construct regulation and support the regular reconstruction of organisational rules. Providing resources – especially time – to construct such ways of handling variability is one recommendation that could be made to companies. Reducing the power of support functions and restoring it to field managers is another. This would mean allowing managers to spend time saying things that could be potentially career threatening.

References

Cru, D. (2014). *Le risque et la règle. Le cas du bâtiment et des travaux publics*, Toulouse: Erès.

Dejours, C. (1980). *Travail: usure mentale*, republished, Paris: Bayard, 1993.

Gauthereau, V. and Hollnagel, E. (2005). Planning, control and adaptation. *European Management Journal, 23*(1), 118–131.

Gordon, T., Lahelma, E., Hynninen, P., Metso, T. Palmu, T. and Tolonen, T. (1999). Learning the routines: "professionalization" of newcomers in secondary school, *Qualitative Studies in Education, 12*(6), 689–705.

Hayes, J. (2014). The role of professionals in managing technological hazards. The Montara blowout. In S. Lockie, D.A. Sonnenfeld and D.R. Fisher (Eds.), *Routledge International Handbook of Social and Environmental Change*. London & New York: Routledge.

Johansen, J.P., Almklov, P.G. & Mohamad, A.B. (2016). What can possibly go wrong? Anticipatory work in space operations. *Cognition, Technology and Work, 18*(2), 333–350.

Neal, M. and Morgan, J. (2000). The professionalization of everyone? A comparative study of the development of the professions in the United Kingdom and Germany. *European Sociological Review, 16*(1), 9–26.

Wilensky, H.L. (1964). The professionalization of everyone? *American Journal of Sociology, 70* (2), 137–158.

Chapter 4
Captain Kirk, Managers and the Professionalization of Safety

Hervé Laroche

Abstract Historically, management as a means for governing business organizations has developed at the expense of professions as autonomous, self-regulated bodies. Therefore, the current call for "professionalization" in the domain of safety might be surprising. This chapter explores this apparent contradiction in the form of an imaginary dialogue between an operator and a manager. The current "injunction to professionalism" is critically assessed. Alternative views of professionalization are developed, with implications for alternative managerial roles.

Keywords Managerialization · Accountability · Empowerment

O: Operator
M: Manager
M: The company will shortly launch a training program for improving safety.
O: Another boring series of sessions with a guy who knows nothing about my job. More stupid slides that would make a five year old cry.
M: I know, you've had too much of that. This time it's different. It's not only about rules, more about a "professionalization of safety".
O: Professionalization?
M: Yes. Precisely, not taking you as a five year old but rather as a professional who should know what he's doing and why.

This chapter takes the unusual form of a dialogue between an operator and his manager in an industrial company (with a concern with safety). It deliberately takes a critical stance towards the idea of professionalization. This does not reflect my entire views on the topic. In fact, I share many ideas that are developed elsewhere in the book and that do not need to be reiterated. The critical ideas expressed by the operator and his manager do, however, reflect some of my opinions, though I push them to the extreme for the purpose of debate and also for fun.

H. Laroche (✉)
ESCP-Europe, Paris, France
e-mail: laroche@escpeurope.eu

C. Bieder et al. (eds.), *Beyond Safety Training*, Safety Management,
https://doi.org/10.1007/978-3-319-65527-7_4

O: I know what I'm doing and why. More than you do.

M: Let's not argue.

O: What you really want is for me to do what you want me to do, without you having to tell me.

M: It's not only me. My bosses. The customer. The HSE department. The inspectors. You know.

O: Lots of people with lots of different ideas about how I should do the job. And everybody trying to cover their own ass.

M: Please.

O: Look, I know you're doing your best. But when I hear that kind of stuff about "professionalization of safety" or whatever, I find it hard to take it seriously.

M: Why?

O: Since I've been working here, there have been more rules and norms every day. Plus auditing, reporting, paperwork. Nobody wants me to be a professional. Everybody wants me to comply with things nobody asked me about before. Including stupid rules and pointless norms that prevent me from doing a better job. You know very well that I could do a much better job. And a safer one, too.

M: That's what this program is about!

O: Really? You're going to let us operators organize the work?

M: Well… not quite. But I promise you, we'll discuss any issue you'll raise.

O: See? You managers don't want professionals. You never did. More than that: management is all about substituting abstract rules and norms for tacit knowledge. It's about replacing self-organizing workers with individuals complying to standards and orders. It's been like that since good old Winslow E. Taylor.

M: This is not like that anymore.

O: Yes it is! More and more! You're even doing this to the eggheads now. In hospitals, universities, law firms. Extracting and commodifying their knowledge, restricting autonomy, evaluating through your own standards of quality and productivity. Of course, you have to accept some degrees of autonomy. After all they're doing the real job, you need them. But every bit of power you can take away from them, you take it.

M: How do you know about all this?

O: I read the Sunday papers. The truth is, managerialization is just incompatible with professionalization. The drive for professionalization is normally from within a community of equals. They organize themselves to gain credit and define their own territory. Sociologists call this a jurisdiction (Abbott 2014).

M: Why is it different in our case?

O: Well, the call for professionalization comes from above. It's a top-down injunction to professionalize (Boussard 2009). An oxymoron, to some point. Though not exceptional. Managerialization often comes with a discourse about professionalization. This is strange, when you think of it. That's why I'm suspicious.

M: You forget that managers have to demonstrate that the organization complies with external norms and standards. Not mentioning the expectations of customers, governments, the press, and the public at large.

O: You're right. The injunction for professionalism comes also from outside. It's passed on to the managers.

M: Funny, when I was in India it was the workforce who called for more professionalism from their managers! Mostly they were speaking out against nepotism, abuses, harassment, and a general lack of managerial skills. You see, there's also pressure from below (Vaidyanathan 2012).

O: Thanks for the idea. We should do that to.

(*They laugh.*)

M: So why would the company top management launch a program called "professionalizing safety"?

O: They're clever guys. They sell old stuff in a new bottle. Good idea, really. Everybody wants the guy who handles the risky stuff to be a "professional", whatever this means. And among the workforce, who can oppose a professionalizing program? Some of my worker buddies might even fall for it.

M: Well isn't it truly appealing?

O: I'm not buying. It's a trick. It's patting you on the shoulder. It's chocolate medals. Some guys, they called this "grandiosity" (Alvesson and Gabriel 2015). This is another thing you managers are very good at. Calling things by a fancy name. Take Mission Statements for instance. Dedication. Commitment. Serving the community. Social responsibility. Company culture. Professionalism.

M: OK, you have a point. I resent that too.

O: I know. Otherwise I would not even talk to you.

M: See, they want me to be a leader, not just a manager. We had a training about leadership skills, you know. We had this lecture about great leaders, Alexander, Julius Caesar, Churchill... Even Napoleon!

(*They laugh.*)

O: So you feel better as a leader?

M: No. (*Sighing*) Sometimes I feel like a subordinate on the Enterprise...

O: You mean, the starship in the Star Trek series?

M: Yes. You have this Captain Kirk, really quite a nice guy, saying "Make it so" whenever he has reached a decision. In the series, you never see the poor guy who actually "makes it so". That's me. I have to "make it so" without bothering Captain Kirk with the details. This is what I'm paid for. If I escalate a problem my boss quickly reminds me about that.

O: They just want to mind their business without hearing from us down below.

M: Most of the time they just don't want to hear about what's wrong. Especially if it's really a tricky issue and if it does not align with their objectives, policies, and all. One thing they really hate is being confronted with their own contradictions and their powerlessness, because of a lack of skills or knowledge or budget or resources or influence or courage or whatever.

O: That's the point. That's why they come up with this professionalization idea. They think that if we're better trained, if we have better skills, they won't hear from us, safety-wise. Because professionals don't complain, they solve issues at their level. That's why they want professionals. Only they don't really want professionals. They just don't want problems coming to their attention.

M: You're probably right…

O: And in the end, me and my work buddies end up with the nasty details. Like having to violate safety rules for the purpose of productivity.

M: Yes. I try to avoid it but most of the time I can't do otherwise.

O: That's not new. This is how big companies work. It's called "pushing down the details" (Jackall 1988). It hides the fact that managers often don't know how to solve issues or don't want to solve them because it would acknowledge that wrong choices and inadequate policies have been implemented. It hides contradictions within the organization. "Professionalism" boils down to a general injunction to handle the details. It's the same "Make it so!", only wrapped within an abstract, supposedly appealing discourse of special skills and abilities.

M: Yet there might be something to take from it. What would that be, then, truly professionalizing safety?

O: I'll tell you a story (Hampden-Turner 1990). In another life, I was a truck driver for an oil and gas company in the Rocky Mountains. Yes, the wild, wild West. And the truck I was driving was a tanker. We had bonuses and penalties for delivering to the gas stations in time. We were proud of being professional truck drivers. Safety was not a priority. Violations when loading or unloading were the rule. Speed limits on the highway and through the small towns were optional. We did everything to speed up the process. Management had tried everything. Big books of rules that nobody read. Inspectors that everybody knew how to fool. Endless training sessions that sounded like Sunday school. One day the guy who did the training was so rudely pushed around that he ran away in tears and filed a complaint. The drivers didn't care. They thought that risk was part of the job and that they should take it like a man takes life: in his own hands.

M: Like a shotgun.

O: Exactly. One day an elderly guy came and called for a meeting. We thought it was another silly training session. But the guy said he had nothing to teach us about safety, only that we did not behave safely. He said that parents were afraid that their kids would get run over when a tanker rushed by their homes. He said that gas station attendants and all their neighborhood feared that one day a tanker might blow up when unloading. He said that we were a hazard for everybody in the country. Then he listened to us. We said that the schedules were too tight; that many safety procedures were stupid; that the trucks themselves were poorly equipped; that the roads were dangerous; that we had to make a living. We shouted at him but we were ashamed. Then he left. He came back a week later with some guys from the State Road Service, an expert from the technical department, a guy from a truck company, and a couple of gas attendants. We formed committees of volunteers on a variety of themes. We held meetings in towns with the population. We proposed roadwork projects, design ideas for safety devices, a system for establishing delivery schedules, principles for bonuses and penalties, and we participated in the writing of all safety procedures. We committed to reporting violations to an elected group among us. They decided on the sanctions. We still saw ourselves proudly as professional truck drivers. We still thought that risk was part of the job and that we should take it like a man takes life: in his own hands. Only this had a different meaning.

M: Nice story. You should sell it to Hollywood. *(Imitating an advertisement)* Now a major movie, starring Kris Kristofferson!

O: That's what professionalism means. Not only training, not only safety. Empowerment and autonomy. For that managers have to relinquish a lot of their own power. There is no way such empowerment can be limited to specific aspects of the work, such as safety. Safety is a way of doing the work, not an add-on to the tasks. So, empowerment has to encompass the work as a whole.

M: OK... Do I still have a job?

O: Sure. It depends on what you managers want for yourselves. You can be the guy who came to speak to the drivers. You can actively participate in the activities of professionalization. For instance, you can initiate, organize, validate the processes through which the norms and practices are put into coherence by the operators. In doing so, you managers could retain some control over the empowering process and its outcomes.

M: True, but I would have to advocate for substantial changes in norms and practices that can be difficult to accept for upper management levels and/or for outside authorities or stakeholders. And I would be held accountable for the resulting norms and practices.

O: Right. But isn't that what you're paid for?

M: Not that trick again!

O: If you don't like that you can keep your distance with us operators. Operators would have to adapt practices to real-world constraints and find out the best possible trade-offs. As professionals, they would have to regulate themselves so that errors, accidents and violations are kept to a minimum. Or at the very least, they should see that these unwanted events are confined within a restricted area of confidentiality. You managers, in turn, would take charge of the various stakeholders. Your job would be to dress up a convincing window of compliance that would keep authorities and stakeholders satisfied. In short, you would have to erect a protective barrier in order to enable autonomous operators to efficiently do their job. Occasionally you would have to cover up errors and violations (at least what could be identified at as such by an outsider), as long as the real stuff is taken care of.

M: The obvious limitation of this strategy is that things can get out of control. In case of a major failure, exposure will be maximum.

O: Yes. The difference with "true" professions like health specialists is that lower-level operators will never be held fully accountable. In case of a major failure, management will take the blame (which is only normal).

M: I'm not keen. Unfortunately, safety is not the only outcome at stake. As autonomy cannot be solely safety-related, the more general issues of costs and efficiency of the workforce come into play.

O: I know. There is probably a range of intermediate strategies. Yet the basic principle here is an explicit (though not advertised) trade-off between autonomy for the operators and accountability for management, with the idea that it is for the better. The more managers really believe that operators' autonomy will lead to better safety, lesser failures, and subsequently less exposure for themselves, the more they may accept a clear division of labor between operators (taking care of the real stuff) and themselves (taking care of stakeholders).

M: Did you think out all this by yourself?

O: Not really. Such a division of labor has been coined the "organization of hypocrisy" by a Swede (Brunsson 1993).

M: You read more than the newspapers.

O: I read a bit of organization studies literature on the week-ends.

M: While I watch Star Trek again and again!

(*They laugh.*)

References

Abbott, A. (2014). *The system of professions: An essay on the division of expert labor.* University of Chicago Press.

Alvesson, M., & Gabriel, Y. (2015). Grandiosity in contemporary management and education. *Management Learning, 47*(4), 464–473.

Boussard, V. D. (2009). *L'injonction au professionnalisme. Analyses d'une dynamique plurielle.* Rennes, France: Presses Universitaires de Rennes.

Brunsson, N. (1993). Ideas and actions: Justification and hypocrisy as alternatives to control. *Accounting, Organizations and Society, 18*(6), 489–506.

Hampden-Turner, C. (1990). *Corporate culture: From vicious to virtuous circles.* London, England: Hutchinson Business Books.

Jackall, R. (1988). *Moral Mazes. The world of corporate managers.* New York: Oxford University Press.

Vaidyanathan, B. (2012). Professionalism 'from below': mobilization potential in Indian call centres. *Work, Employment & Society, 26*(2), 211–227.

Chapter 5
A Critique from Pierre-Arnaud Delattre

Pierre-Arnaud Delattre

Abstract In this chapter, the author mainly addresses the differences between France and Anglo-Saxon countries regarding two axes. First he shows the differences in terminology of the word 'professional' and related terms, then he highlights that their respective approaches of human and organisational factors in Occupational Health & Safety originate from their own specific history.

Keywords OH&S · Professional · Human factor

Overall, the FonCSI seminar held on 12 November 2015 in Chantilly[1] was well structured and organised. The speakers were captivating and the topics presented were relevant to today's industrial challenges. It was rewarding to be a part of rich and mature exchanges on subjects that have real and practical implications in the day-to-day management of industrial OH&S.[2]

Discussions around definitions, particularly by Professor Rhona Flin on the use of the term "professional" when referring to the safety function, highlighted that on an international stage, we may sometimes assume we speak the same language of OH&S but in fact we would benefit from defining in more detail some of the terminology we freely use in passionate conversations on our vocational topics. This aspect of the seminar was, for me, one of the richest contribution; that experts ("professionals") in the field of OH&S, but belonging to different national cultures, could hold discussions with so much in common, and yet a handful of critical definitions could make such a big difference in perceptions, understanding and even connectivity between the parties holding the discussions. In the world of multi-national organisations and particularly with respect to OH&S training, this

---

[1]The two-day international workshop mentioned in the preface and highlight of the project that led to this book (editors' note).
[2]Occupational Health & Safety (editors' note).

P.-A. Delattre (✉)
Value 4 Life, Perth, Australia
e-mail: pdelattre@value4life.com.au

C. Bieder et al. (eds.), *Beyond Safety Training*, Safety Management,
https://doi.org/10.1007/978-3-319-65527-7_5

41

may be an important aspect of communication we overlook. The practical simulation training presented by Professor Vincent Boccara was a positive and refreshing approach to OH&S training which would overcome communication barriers like these, as well as those self-imposed between hierarchical levels of an organisation (e.g. Supervision and Management or Workers and Supervision).

Based on observations made while studying Industrial Risk at La Sorbonne University, following a period of twelve years working in an English setting within a Swiss Company, I propose, as a thesis for this paper, that there have been apparently different approaches to OH&S adopted by experts from the English world and the French world.

At the risk of over-simplifying, it is my belief that the English approach to OH&S, greatly influenced by a number of historical catastrophes such as Piper Alpha, was originally driven by structured organisational processes and systems, having matured from a prescriptive (regulator driven) to a risk-based approach post-Cullen enquiry. This called for organisations to demonstrate, through OH&S experts and through Safety Cases, that the risks in work tasks had been mitigated as much as was reasonably practicable. Until more recently, the human in this approach, had been largely passive apart from a requirement to apply the training received. Human factor engineering and, more recently, behavioural psychology is offering the current direction for development so that the paradox of human excellence and fallibility may be taken into account in the design of Work, in particular when involving machines and equipment and moreover when the work tasks affect other human beings; for example, in the mass transportation, restauration or medical fields, because of the potentially catastrophic consequences of an incident.

The French Approach has evolved from the Napoleonic era, when the labour code was created to protect the children and women working and dying in great numbers in chemical and mechanical factories, while the men were either working the land in the fields or at war. It was driven by an assessment of the physical limitation of humans and the subsequent adaptation of the work tasks to those humans by engineers, and has now evolved into the scientific (physical, mental, psychological) human factors engineering where the French and the English approaches finally meet.

The FonCSI seminar on 12 November was, for me, a fantastic modern meeting place with additional views from Italy, Netherlands, Belgium and Norway. The potential to share such international experience can only help to create improvement by combining our approaches to OH&S, shared and illustrated through diverse lessons learnt from industry and academia.

This leaves one question for future events: are we casting the net wide enough to learn from and engage with the emerging industrial nations in the Middle East, India, Asia, Africa and South America?—language permitting!

Chapter 6
Enhancing Safety Performance: Non-technical Skills and a Modicum of Chronic Unease

Rhona Flin

Abstract Current debates on professionalism and safety cover a range of inter-pretative challenges and theoretical perspectives, as the workshop organized by FonCSI in 2015 revealed. One avenue for consideration was to address the question of the role of professionalism in the job with regard to safety. For example, should safety training just be part of normal job training or should it have a separate and distinctive position in the training curriculum? In this paper, I consider two ways in which safety training and safety thinking are being integrated into routine managerial and technical work. The first of these is behavioural, namely to focus on the non-technical skills (NTS) for a given job, as evidenced by the airlines' Crew Resource Management training and assessment programmes. This approach is now being adopted in other safety-critical sectors, such as acute medicine and offshore oil and gas operations. The second direction is more attitudinal in nature: it examines the relatively novel concept of chronic unease, derived from the High Reliability Organisation literature. These two approaches show that addressing both workplace behaviours (non-technical skills) and underlying attitudes to operational risks (chronic unease), can help to build protective skills for safety into the professional job repertoire.

Keywords Professionalism · Crew Resource Management · Chronic unease

R. Flin (✉)
Robert Gordon University, Aberdeen, UK
e-mail: r.flin@rgu.ac.uk

© The Author(s) 2018
C. Bieder et al. (eds.), *Beyond Safety Training*, Safety Management,
https://doi.org/10.1007/978-3-319-65527-7_6

6.1 Introduction

The opening position statement for the FonCSI workshop[1] (see Introduction by
Gilbert) came from a concern of the member companies, namely that training
programmes in the field of industrial safety no longer appear to be yielding the
expected results. This contribution is primarily directed at questions proposed for
the FonCSI workshop on professionalism and safety

> What part could professionalism in the job play in safety? Should safety training be
> incorporated into everyday practices and activities or should there be specific safety
> training? (Foncsi 2015)

The reason for choosing this topic is that much of my research has been founded,
perhaps implicitly at times, on the assumption that enhancing job performance so
that it is of better quality and efficiency will concomitantly enhance safety due to
improved risk perception and risk management behaviours.

In this chapter, I first briefly discuss what I understand the term 'professionalism'
to mean. It is not a topic I have ever studied and so I have attempted to set out my
interpretation of how professionalism relates to workplace safety. I then discuss two
areas where my own safety research has been located. These indicate two ways in
which safety training and safety thinking can be integrated into routine work, at
both operator and managerial levels. The first is behavioural, namely to focus on the
non-technical skills for a given job, as evidenced by the airlines' Crew Resource
Management training and assessment programmes. This approach is now being
adopted in other safety-critical sectors, such as acute medicine and offshore oil and
gas operations, typical for operational staff but in some cases also for managers. The
second direction is more attitudinal in nature: it examines the relatively novel
concept of chronic unease, derived from the High Reliability Organisation literature. This has been used at both operational and managerial levels. What is proposed is that addressing both non-technical skills, as well as underlying attitudes to
operational risks (chronic unease), can help to build protective skills for safety into
the professional job repertoire.

6.2 What Is Professionalism?

What does 'professionalism' in the job mean? Is it about having defined standards,
specified programmes of education and qualification monitored by subject matter
experts? The term 'profession' has a long history, traditionally referring to specialist
occupations based on an extensive body of knowledge, such as law, divinity or
medicine which have controlled qualifications and specified training leading to

---

[1]The two-day international workshop mentioned in the preface, organized by FonCSI in November
2015 and highlight of the project that led to this book (editors' note).

membership of the professional body. Professionalism is a newer conceptualisation, reflecting an extended range of occupations now seen as professions and the idea of 'professional standards/professional behaviour' being seen as part of many modern jobs. Sociologists have engaged in extended debate about professionalism versus managerialism, normative versus ideological interpretations, the rise of professionalism and its implications for organisational life (Evetts 2003; Noordegraaf 2011). In relation to safety, we can distinguish between:

1. the increasing professionalization of the safety specialist, in the form of the safety adviser or safety manager and
2. embodying an additional focus on safety into the skills of the technical professional.

In relation to the first point, it is important to recognise that the safety specialists have a valuable role in many organisations, especially with regard to regulatory compliance, large scale audit, and design and implementation of safety management systems. My own work has been more concerned with the second approach. Namely, trying to identify how an appropriate skill set for enhancing safety can be identified so that this can be incorporated into professional development (whether technical or managerial).

6.3 Crew Resource Management and Non-technical Skills

One of the most obvious demonstrations of this approach of trying to build safety skills into general professional competence is Crew Resource Management (CRM). This is a training approach introduced by the aviation industry in the 1980s, following the realisation that a focus on technical skills was not sufficient. Accident analyses, which greatly benefitted from cockpit voice recorders, showed clearly that deficiencies in teamwork, leadership, decision making, situation awareness and communication were contributing to adverse events (Kanki et al. 2010). Of course, this was not to say that organisational factors, managerial behaviours, company culture and work conditions were not also exerting a powerful influence on airworthiness, technical reliability and flightdeck behaviours (Maurino et al. 1995). Notwithstanding these powerful influences on the behaviours of front-line staff, the personnel closest to the hazards may have the opportunity to enhance or diminish the level of flight safety by their actions. Their behaviour can increase exposure to risk (e.g. rule violation) or can be protective (e.g. by monitoring, and if necessary challenging, the actions of other crew members).

Task analyses using interviews, surveys, simulator observations and accident analyses were employed to identify pilots' CRM skills, which were essentially protective for safety, by reducing the incidence of error or by 'catching' or mitigating errors that occurred. Errors jeopardise efficiency, as well as safety. Improved communication enables smoother interaction between crew members and supporting personnel. Awareness of human performance limiting factors, such as stress and fatigue results in better self-monitoring and corrective action. Having determined the principal skill set, classroom and simulator-based training courses, called CRM training, were devised to teach pilots (and subsequently cabin crew and aviation engineers) the basic knowledge of the psychological principles underlying these non-technical skills and to show why the associated behaviours were protective for flight safety.

In European aviation, the regulator (JAA[2] at the time) when discussing Crew Resource Management introduced the term 'non-technical skills', essentially the cognitive and social skills that complemented the pilot's technical skills and enhanced efficiency and safety (Flin et al. 2008). The teaching and assessment of non-technical skills is obligatory for airline pilots in most countries. For example, in the UK, the Kegworth plane crash in 1989 (where *British Midland* pilots mistakenly shut off the working engine when the other was on fire) was such a strong demonstration that human error and teamwork failures were contributing to fatal accidents, that the Civil Aviation Authority (CAA) took the view that CRM had to be introduced, even though at the time there were only a few scientific studies on its effectiveness. In the ensuing years, there have been many advances in CRM training and in 2015 the European Aviation Safety Authority (EASA) released new guidance on Crew Resource Management (EASA 2015), an example of which is shown below.

> CRM training should be conducted in the non-operational environment (classroom and computer-based) and in the operational environment (flight simulation training device (FSTD) and aircraft). Tools such as group discussions, team task analysis, team task simulation and feedback should be used.
>
> CRM principles should be integrated into relevant parts of flight crew training and operations including checklists, briefings, abnormal and emergency procedures.
>
> CRM training should address hazards and risks identified by the operator's management system described in ORO.GEN.200.
>
> Whenever practicable, the compliance-based approach concerning CRM training may be substituted by a competency-based approach such as evidence-based training. In this context, CRM training should be characterised by a performance orientation, with emphasis on standards of performance and their measurement, and the development of training to the specified performance standards.

One of the strengths of the CRM approach is that the training content is based on underlying scientific evidence from psychology, physiology or other relevant disciplines (Kanki et al. 2010). Thus it strives to continually develop and foster evidence-based practice. Another fundamental principle of CRM is that the training content should be designed to address current operational issues and to reflect learning from adverse events and near misses. These are both illustrated in the section below.

---

[2]Joint Aviation Authorities (editors' note).

6.3.1 Startle Effects

One example of this process of incorporating learning from adverse events into CRM training relates to the Air France accident (2009). An Airbus (AF447) crashed into the Atlantic Ocean when flying between Rio de Janeiro and Paris (BEA 2012). One of the contributing factors to the accident was that the pilots had apparently become startled by rapid changes in aircraft state. The EASA (2015, p. 5) guidance, mentioned above, addresses this issue.

(3) Resilience development

CRM training should address the main aspects of resilience development. The training should cover:

(i) Mental flexibility
    Flight crew should be trained to:

(A) understand that mental flexibility is necessary to recognise critical changes;
(B) reflect on their judgement and adjust it to the unique situation;
(C) avoid fixed prejudices and over-reliance on standard solutions; and
(D) remain open to changing assumptions and perceptions.
(ii) Performance adaptation
     Flight crew should be trained to:

(A) mitigate frozen behaviours, overreactions and inappropriate hesitation; and
(B) adjust actions to current conditions.
(4) Surprise and startle effect
    CRM training should address unexpected, unusual and stressful situations. The training should cover:

(i) surprises and startle effects; and
(ii) management of abnormal and emergency situations, including:

(A) the development and maintenance of the capacity to manage crew resources;
(B) the acquisition and maintenance of adequate automatic behavioural responses; and
(C) recognising the loss and re-building situation awareness and control." (p. 5)

These additions to CRM training had to be implemented by European operators by October 2016. This is an excellent example of how fundamental training for a professional group is continually reviewed and revised (in this case by an international regulatory body) to take into account emerging issues relating to safety that have not previously been recognised to this degree.

As mentioned above, research evidence is used to develop the content of CRM training and this has also been true for the startle effect phenomenon. When EASA were reviewing their European CRM guidance, specific research findings on startle effects were sought and considered. In the USA, the Federal Aviation Administration (FAA) commissioned research into this phenomenon (e.g. Rivera et al. 2014) and on the more general effects of acute stress on aircrew performance (Dismukes et al. 2015). There are only a limited number of studies but these have shown individual variation in response and recovery patterns, as well as pilots' awareness of this effect (Martin et al. 2015, 2016). And of course, startle effects are not peculiar to pilots. Recent studies in healthcare indicate similar patterns of reaction to unexpected events with concomitant delays in decision making and responsive action during resuscitation (Lu et al. 2015).

6.3.2 CRM Beyond the Flightdeck

The CRM/non-technical skills approach has now extended into many other occupations, including the mariners and ship engineers; railway workers, miners, systems analysts (Flin et al. 2008, 2014). In healthcare, there has been particular interest from members of operating theatre teams with non-technical skill sets developed for anaesthetists (ANTS); scrub nurses (SPLINTS); anaesthetic practitioners (ANTS-AP) and surgeons (NOTSS). There are now training courses on non-technical skills provided for these occupations and the Royal Australasian College of Surgeons incorporated the NOTSS framework into their new professional standards (Flin et al. 2015). As suggested above, this shows how an increased emphasis on behaviours targeted to improve safety (via a non-technical skills approach) can be adopted as part of existing professional development.

The main objective of this CRM/ NTS training and assessment has always been to improve safety/reduce accidents, hence the behavioural rating scales tools to measure performance on non-technical skills are phrased in the language of safety. For example, in the NOTECHS system for pilots (van Avaermaete and Kruijsen 1998; O'Connor et al. 2002), the scale descriptors use explanatory terms such as

'behaviour directly endangered flight safety' *or* 'behaviour enhances flight safety'

for very poor and good performance respectively. But the behavioural examples (markers) in such systems relate not to specific safety-related activities but to normal task operations. Thus the underlying premise is that better demonstration of skills such as leadership, teamwork, decision making during task execution will benefit safety. The medical professionals who developed non-technical skills frameworks and associated behavioural rating systems (e.g. ANTS, NOTSS) have adopted the same type of scale descriptors (Flin et al. 2015), in this case with the purpose of emphasising that patient safety is of paramount importance.

Following the blowout on the Deepwater Horizon drilling rig (2010) in the Gulf of Mexico that killed 11 workers and injured a further 50, as well as creating an enormous marine pollution event (Report to the President 2011), the offshore oil and gas industry became interested in applying CRM to enhance safety (Flin et al. 2014). Social scientists' analyses of the accident show clearly how failures in non-technical skills could have contributed to the trajectory of this event (Hopkins 2012; Reader and O'Connor 2014; Roberts et al. 2015b). In order to design customised training for drillers and other well control specialists, detailed task analyses are required to pinpoint the most important non-technical skills that can help to

protect the safety of the well (Roberts et al. 2015a). In a number of drilling companies, these skills are being taught alongside the technical skills of well control,
especially where there are simulation facilities available that allow for demonstration and feedback.

The evaluation literature on CRM and safety outcomes is somewhat limited in
aviation (low accident rates offer insufficient outcome data) although there have
been meta-analyses (e.g. O'Connor et al. 2008). The more recent introduction of
non-technical skills/CRM to the world of healthcare means that as a technique it is
being scrutinised by a new level of rigour, given medical professionals' concern
with treatment efficacy and willingness to measure error rates and outcomes. Thus
there is an emerging database of studies examining the relationships between
technical skills, non-technical skills, error and safety or other performance metrics.
These generally indicate positive, if patchy, relationships (Hull et al. 2012) but there
is an emerging message that focussing on improving the non-technical skills
required for both routine and abnormal task activities can improve safety.

In the following section, I consider a second approach that is being adopted by
some companies which is not just to consider behaviours but also to foster particular attitudes or 'mind sets' that will drive the choice of behaviours that should
enhance safety. This is another approach to building safety professionalism into
everyday task activities, and the mental state in question is called 'chronic unease'.

6.4 Chronic Unease

In a work environment that has few accidents, even though there are significant
hazards present, there is a likelihood that the risks are underestimated leading to a
false sense of comfort or complacency. A report by Cass Business School (2011),
investigated major corporate crises (explosions, fires, product-related and supply
chain crises, and IT[3] problems) and identified failures at board level in these
organisations. There was an inability to recognise potential risks and engage with
them, a tendency to ask fewer questions when things were going well and not
recognising changes in the corporate environment.

Recent industrial interest in applying the concept of 'chronic unease' to managerial and operational thinking on safety matters is an attempt to address this type
of problem. The concept comes from the literature on 'high reliability organisations' (HRO). A key HRO characteristic is the lack of complacency about risks. For
instance, with regard to the structural failures that caused the Alexander Keilland
drilling rig accident in Norway, Weick (1987, p. 119) commented

> Part of the mind-set for reliability requires a chronic suspicion that small deviations may
> enlarge, a sensitivity that may encourage a more dynamic view of reliability.

---

[3]Information Technology (editors' note).

The term 'chronic unease' was introduced by Reason (1997) to capture tendencies of wariness towards risks, thus as a contrast to complacency. He described it as resulting from an absence of negative events, leading *'people [to]forget to be afraid'* (p. 39). The Oxford English dictionary defines 'unease' as a form of discomfort and distress, related to strain and representing a feeling of concern.

In a similar vein, the HRO literature discusses alertness and good management of risks under the label of 'mindfulness' in organisations. Weick and Sutcliffe (2006) describe mindful organisations as:

1. dealing with risks by investing substantial resources, both financial and attentional,
2. early detection of issues,
3. pre-occupation with failure,
4. reluctance to simplify,
5. sensitivity to nuances that can lead to failure,
6. commitment to resilience and
7. willingness to defer to experts.

The resilience literature uses the term 'restless mind' to label awareness that things can go wrong and alertness to weak signals (Westrum 2008). Likewise, Pidgeon (2012) writes of 'safety imagination' to describe inadequate appreciation of risk in relation to the Fukushima nuclear power plant accident in Japan in 2011.

My research group had been studying managers' safety leadership and safety commitment (Agnew and Flin 2014; Flin 2006; Fruhen et al. 2014) and we were sponsored by a multinational oil and gas company to work on a study of chronic unease in managers. This company had already developed safety materials for operational staff which emphasised the importance of having a sense of respectful unease for ever-present risks in the work environment and confirmed the importance of vigilance and attention to weak signals. The focus of our project was on how more senior managers might experience chronic unease, how that could influence their behaviour and whether they felt this was beneficial for safety and for the business more generally.

Despite the prevalence of the term 'chronic unease' in the high reliability organisation (HRO) literature, there was limited evidence to enable a definition or operationalisation of this concept. To develop a better understanding of chronic unease, we conducted a literature search using this term (Fruhen et al. 2014). We only found descriptions of chronic unease in 9 articles. These were coded resulting in the identification of five themes: pessimism, propensity to worry, vigilance, requisite imagination and flexible thinking. From the descriptions in the literature and with reference to related conceptualisations, we proposed the following components:

- **Pessimism** is a disposition that drives individuals to anticipate failures and expect negative events and therefore may promote chronic unease about safety. Pessimism is not concerned with emotions and somatic reactions, but rather represents an attitude towards the future.

- **Propensity to worry** is a tendency towards experiencing an emotional reaction with regard to possible failure. This could be characterised as a personality trait (worry regularly about many things) or as a more transient state (to worry about a particular event). It is a subcomponent of anxiety, described as specific to cognition that is characterised by concern about future events.
- **Vigilance** is usually defined in terms of an observer's ability to maintain attentional focus and to remain alert to stimuli over prolonged periods of time. Thus it concerns the detection of signals in the environment.
- **Requisite imagination** was defined by Adamski and Westrum (2003, p. 195)

  the fine art of anticipating what might go wrong.

  It is not primarily concerned with the exploration of current states of problems, but with the ability to project their future development.

- **Flexible thinking** relates to creative problem solving which typically involves divergent thinking. We suggested that chronic unease promotes this mode of cognition.

We proposed a preliminary conceptualisation of chronic unease based on these attributes, arguing that this particular mental state may be desirable for managers in relation to the control of risks. Desirable because the feelings of unease may lead them to be more attentive to risk information and to incorporate a safety dimension into their operational decision making.

In a second study, we carried out semi-structured interviews with 27 senior managers from several companies in the energy sector (Fruhen and Flin 2016). The aim of the study was to determine if the five components identified in the literature review would be evident in the managers' responses when discussing their safety leadership practices. We were also interested in how a sense of chronic unease would affect the managers' behaviours. Content analysis of the interview transcripts identified flexible thinking most frequently, followed by pessimism, propensity to worry, vigilance and requisite imagination. Flexible thinking was frequently also coded as a behaviour, suggesting it to be a partially observable response to chronic unease. Other behaviours that emerged as related to chronic unease were demonstrating safety commitment, transformational and transactional leadership styles, and seeking information. Chronic unease was described as having positive effects on safety, positive and negative effects on team interaction and negative effects on business and the managers' personal outcomes. We concluded that the five components provide a basis for the measurement of chronic unease and suggested behaviours and responses that should be considered in its future investigation. Figure 6.1 provides a proposed model of how the components and associated behaviours might be related.

In terms of increasing the professionalization of safety in managers, it is suggested that attention should be paid to these underlying characteristics and attitudinal states ('mind-set for reliability'), as well as considering the related behaviours. The extent to whether 'chronic unease' in managers can be developed is still to be

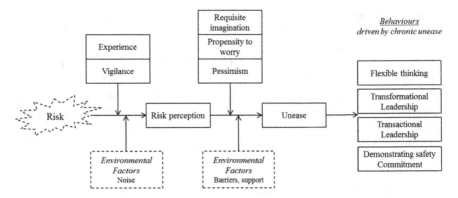

Fig. 6.1 Proposed model of the influence of chronic unease on managers' behaviours (Fruhen and Flin 2016). Reprinted from the *Journal of Risk Research* with the permission of Taylor and Francis

determined. Managerial safety training which endeavours to increase risk aware-ness, to demonstrate the personal consequences for managers of having a major accident and to show how companies can become complacent is essentially trying to increase a sense of unease.

When introducing the topic of chronic unease, I am often asked—'So how much chronic unease is desirable?' Certainly a high level of chronic unease for safety could have negative consequences for individuals who are consumed by worry and constantly fearful of catastrophic events on their worksites. Of course, any extreme mental state can be harmful; at an abnormal level there could be reduced well-being, stress and even clinical anxiety and depression. Thus we proposed that the relationship between chronic unease and efficacy in safety management prob-ably has a curvilinear nature (Flin and Fruhen 2015). Too little unease and the resulting complacency could mean that warning signals are ignored, ambiguities are marginalized, there is no systematic search for negative indicators, and adverse consequences are rarely considered. Too much unease and the manager could be disabled by anxiety with consequent deleterious effects on decision making, action and mental health. At the optimal level, which will be individually determined, the sense of chronic unease about organisational safety prompts a continued search for hidden threats, the extended consideration of ambiguities and anomalies, and the appreciation of disconfirming evidence.

This suggestion of a curvilinear relationship between for chronic unease and managerial performance is based on Janis and Mann's (1977) conflict model of stress and decision making, where they described coping patterns in decision conflict situations (i.e. ambiguity about the best option) with distinctive levels of stress. They describe various states including 'hypervigilance' where there is recognition of the serious risks in the alternative courses of action. In this case the stress level is extreme (cf. high chronic unease) creating a state akin to panic with the individual preoccupied with the threatened losses. Resulting behaviours can include impulsive actions, vacillation and simplistic, repetitive thinking. We argued

(Flin and Fruhen 2015) that such high levels of anxiety might exist because a manager is not well-suited to the job, or has insufficient knowledge and support. Alternatively, such high anxiety could be a strong indication that there are uncontrolled risks and that continuing with workplace operations could result in an accident.

It was suggested that the desirable level of chronic unease for safety is in the mid-range, perhaps this could be called 'a modicum of unease'. This is similar to what Janis and Mann called vigilance which was associated with a moderate level of stress. In this case, they described the decision maker recognizing that there are serious risks imbued in competing alternatives, but having confidence about the likelihood of finding an adequate solution in the available time. Clearly there would be marked individual differences in what level of unease would produce discomfort in managers and the types of risks that would most concern them. The professional implications of chronic unease may be more pertinent to organisational selection practices than to safety training.

6.5 Conclusion

As Gilbert (see Introduction) points out in his opposition statement regarding the ineffectiveness of safety training, there is a need to make a finer grained assessment of exactly where there appears to be no or minimal return on investment. It is likely that many safety training courses such as on hazard awareness, use of protective techniques, event analysis, risk protection measures, do deliver the anticipated developments in knowledge, skills and attitudes. Safety training courses in CRM and non-technical skills can also broaden the range of risk awareness, especially to show how social or intrapersonal factors can increase the risk level during task execution. These courses can also focus on specific methods for altering behaviours, for example, relating to speaking up, listening, conducting handovers, problem solving, task briefing which can have longer term effects on hazard awareness levels, as well as shifting norms of acceptable behaviour to improve the safety culture. However, the evidence on effectiveness for both technical and non-technical safety training can be limited and one component of increasing the professionalism of safety may be a requirement for organisations to spend more time and money evaluating the longer term impact of their safety training and other safety interventions. This would require baseline data to be gathered on knowledge, skills and attitudes prior to the training programme and more use of randomised control designs to enable a robust assessment of treatment effects. To enhance the professionalism of safety, whether by developing safety professionals or increasing safety awareness and related skills of operational staff, a strong evidence base is required to determine what should be trained and how the skills can be developed and maintained. In addition, more attention may need to be devoted to personal attributes, such as having a sense of chronic unease when dealing with hazardous activities. In summary, addressing both workplace, on-task behaviours (non-technical skills), as well as underlying attitudes to operational risks (chronic unease), can help to build protective skills for safety into the professional job repertoire.

References

Adamski, A. & Westrum, R. (2003). Requisite imagination: The fine art of anticipating what might go wrong. In E. Hollnagel (Ed.), *Handbook of Cognitive Task Design* (pp. 193–220). Mahwah, NJ: Lawrence Erlbaum.

Agnew, C. & Flin, R. (2014). Senior charge nurses' leadership behaviours in relation to hospital ward safety: A mixed method study. *International Journal of Nursing Studies, 51*, 768–780.

BEA. (2012). *Final Report. Flight AF 447 on 1st June 2009 A330–203*. Paris: Bureau d'enquêtes et d'analyses pour la sécurité de l'aviation civile.

Cass Business School. (2011). *Roads to Ruin: A Study of Major Risk Events: Their Origins, Impact and Implications*. London: Airmic.

Dismukes, K., Goldsmith, T. & Kochan, J. (2015). *Effects of acute stress on aircrew performance: Literature review and analysis of operational aspects*. NASA Report: NASA/TM-2015–218930.

EASA. (2015). *AMC1 ORO.FC.115 Crew resource management (CRM) training. Annex II to Decision 2015/022/R*. Cologne: author.

Evetts, J. (2003). The sociological analysis of professionalism. *International Sociology, 18*, 295–415.

Flin, R. (2006). Erosion of managerial resilience: Vasa to NASA. In E. Hollnagel, D. Woods & N. Leveson (Eds.), *Resilience Engineering*. Aldershot: Ashgate.

Flin, R. & Fruhen, L. (2015). Managing safety: Ambiguous information and chronic unease. *Journal of Contingencies and Crisis Management, 23*, 84–89.

Flin, R., O'Connor, P. & Crichton, M. (2008). *Safety at the Sharp End. A Guide to Non-Technical Skills*. Aldershot: Ashgate.

Flin, R. Wilkinson, J. & Agnew, C. (2014). *Wells Operations Crew Resource Management*. International Oil and Gas Producers Report 501. London: IOGP.

Flin, R., Youngson, G. & Yule, S. (2015). *Enhancing Surgical Performance. A Primer on Non-Technical Skills*. London: CRC Press.

Foncsi. (2015). Is professionalization a safety issue…or the other way around? Strategic analysis call for papers. Foundation for an industrial safety culture. https://www.foncsi.org/en/research/research-themes/professionalisation-folder.

Fruhen, L., Flin, R. & McLeod, R. (2014). Chronic unease for safety in managers: a conceptualisation. *Journal of Risk Research, 17*(8), 969–979.

Fruhen, L., Mearns, K., Flin, R. & Kirwan, B. (2014). Skills, knowledge and senior managers' demonstrations of safety commitment. *Safety Science, 69*, 29–36.

Fruhen, L. & Flin, R. (2016). 'Chronic unease' for safety in senior managers: an interview study of its components, behaviours and consequences. *Journal of Risk Research, 19*, 645–663.

Hopkins, A. (2012). *Disastrous Decisions*. Sydney: CCH Australia.

Hull, L., Arora, S., Aggarwal, R., Darzi, A., Vincent, C., & Sevdalis, N. (2012). The impact of nontechnical skills on technical performance in surgery: a systematic review. *Journal of the American College of Surgeons, 214*, 214–230.

Janis, I. & Mann, L. (1977). *Decision Making: A Psychological Analysis of Conflict, Choice and Commitment*. New York: Free Press.

Kanki, B., Helmreich, R. & Anca, J. (2010) (Eds.). *Crew Resource Management (2nd edition)*. San Diego: Academic Press.

Lu, S., Waller, M., Kaplan, S., Watson, A., Jones, M. & Wessel, D. (2015). Cardiac resuscitation events: One eyewitness is not enough. *Pediatric Critical Care Medicine, 16*, 335–342.

Martin, W., Murray, S., Bates, P. & Lee, P. (2015). Fear-potentiated startle: A review from an aviation perspective. *International Journal of Aviation Psychology, 25*, 97–107.

Martin, W., Murray, P., Bates, P. & Lee, P. (2016). A flight simulator study of the impairment effects of startle on pilots during unexpected critical events. *Aviation Psychology and Applied Human Factors, 6*, 24–32.

Maurino, D., Reason, J., Johnston, N. & Lee, R. (1995) *Beyond Aviation Human Factors*. Aldershot: Ashgate.

Noordegraaf, M. (2011) Risky business. How professionals and professional fields (must) deal with organizational issues. *Organization Studies, 32*, 1349–1371.

O'Connor, P., Hörmann, H. J., Flin, R., Lodge, M., Goeters, K. M., & JARTEL Group, T. (2002). Developing a method for evaluating Crew Resource Management skills: A European perspective. *The International Journal of Aviation Psychology, 12*, 263–285.

O'Connor, P., Campbell, J., Newon, J., Melton, J. Salas, T & Wilson, K. (2008). Developing a method for evaluating Crew Resource Management skills: a European perspective. *International Journal of Aviation Psychology 18*, 353–368.

Pidgeon, N. (2012). Complex organizational failures: Culture, high reliability, and lessons from Fukushima. *The Bridge, 42*, 17–22.

Report to the President. (2011). *National Commission on the Deepwater Horizon Oil Spill and Offshore Drilling*. Washington, DC: US Government Printing.

Reader, T., & O'Connor, P. (2014). The Deepwater Horizon explosion: nontechnical skills, safety culture, and system complexity. *Journal of Risk Research, 17*, 405–424.

Reason, J. (1997). *Managing the Risks of Organizational Accidents*. Aldershot: Ashgate.

Rivera, J., Talone, A., Boesser, C., Jentsch, F. & Yeh, M. (2014). Startle and surprise on the flight deck: Similarities, differences and prevalence. In *Proceedings of the Human Factors and Ergonomics Society Annual Meeting* (Vol. 58, No. 1, pp. 1047–1051). Los Angeles, CA: SAGE Publications.

Roberts, R.C., Flin, R., & Cleland, J. (2015a). Staying in the zone: Offshore drillers' situation awareness. *Human Factors, 57*, 573–590.

Roberts, R., Flin, R. & Cleland, J. (2015b). 'Everything was fine' An analysis of the drill crew's situation awareness on Deepwater Horizon. *Journal of Loss Prevention in the Process Industries, 38*, 87–100.

Weick, K. (1987). Organizational culture as a source of high reliability. *California Management Review, 24*, 112–127.

Weick, K. & Sutcliffe, K. (2006). Mindfulness and the quality of organizational attention. *Organizational Science, 17*, 514–524.

Westrum, R. (2008). Resilience and restlessness. In E. Hollnagel, C. Nemeth & S. Dekker (Eds.). *Remaining Sensitive to the Possibility of Failure*. 1–2. Aldershot: Ashgate.

van Avermaete, J. & Kruijsen, E. (1998) (Eds.). *NOTECHS. The Evaluation of Non-Technical Skills of Multi-Pilot Aircrew in Relation to the JAR-FCL Requirements*. Final Report NLR-CR-98443. Amsterdam: National Aerospace Laboratory (NLR).

Chapter 7
Situated Practice and Safety as Objects of Management

Petter G. Almklov

Abstract This chapter focuses on the relationship between representations of work (rules, procedures, models, specifications, plans) and work as a situated practice, performed by real people in always unique contexts. Empirically, it is organized around two main examples, the first one being a discussion of the compartmentalization of safety seen in shipping and the railway sector. It shows how safety, as an object of management, has become decoupled from practice, and how current discourses about safety disempower practitioners and subordinate their perspectives to more "theoretical" positions. The second is based on a study of control room operators in a space research operations setting. Here safety in the sense of avoiding harm to people is not the main concern; rather it is the reliability and robustness of an experiment on the International Space Station that is at stake. This example serves as a starting point for discussing how the research and theory on industrial safety should address the different temporalities of different work situations. It also helps to discuss the role of rules and procedures to support safety, reliability and resilience within the field of safety science. Finally, some propositions about the relationship between situated practice and the management of safety are provided: how invisible aspects of situated work might be important for safety yet hard to manage, how procedures and rules might be integrated parts of situated work as much as representations of it and how different temporalities of work situations should be included in the theorizing of safety and resilience.

Keywords Reliability · Systemic accident · Situated practice

P.G. Almklov (✉)
NTNU Social Research, Trondheim, Norway
e-mail: petter.almklov@gmail.com

© The Author(s) 2018
C. Bieder et al. (eds.), *Beyond Safety Training*, Safety Management,
https://doi.org/10.1007/978-3-319-65527-7_7

59

7.1 Introduction

Safety is a word we use to refer to a state or a condition, not an event in itself. This doesn't mean that nothing happens in a safe condition. On the contrary, safety more often than not depends on practice, on continuous actions and situational adjustments. But these are not in themselves safety. Thus, regulating, managing and controlling safety is always a matter of indirect measures, directed at other things that might influence safety. This book discusses how the professionalization of safety coupled with the increasing interest in managing safety and of training professionals in it, can influence industrial safety. Together with my colleagues[1] I have studied situated practice in a variety of industrial contexts and based on this I will reflect on the relationship between representations of work, that is, descriptions and prescriptions, and the practice of the professionals involved. Organizational studies generally including, to some extent, safety studies, have a tendency to stereotype work (See Suchman 1995; Barley and Kunda 2001). We often fail to capture the nuances in how work is actually performed, and we draw boxes and arrows and superficial models of "workflow" to represent it. The starting point and analytical lens of my discussion is this relationship between representations of work (rules, procedures, models, specifications, plans) and work as a situated practice, something that is performed by real people in always unique contexts.

Empirically, this chapter is organized around two main examples. These are not intended as comparative cases, but as two examples that allow us to develop some ideas about the relationship between professionalization and safety and reliability. The first example is a discussion of the compartmentalization of safety seen in shipping and the railway sector. The key motivation for this part is to show how safety, as an object of management, has become decoupled from practice, and how current discourses about safety disempower practitioners and subordinate their perspectives to more "theoretical" positions. The second is based on a study of control room operators in a space research operations setting. Here, safety in the sense of avoiding harm to people is not the main concern; rather it is the reliability and robustness of an experiment on the International Space Station that is at stake. This example serves as a starting point for discussing how the research and theory on industrial safety should address the different temporalities of different work situations. Secondly, this example invites us to recognize that procedures are (in some cases) an integrated part of situated work (and part of the "distributed cognition" of the control room operators). This serves to elaborate the discussions of the role of rules and procedures to support safety, reliability and resilience within the field of safety science. These examples form the backbone of the chapter but are supplemented with observations from other settings, such as infrastructure and petroleum processing. I conclude by providing some proposals about the relationship between situated practice and the management of safety.

---

[1]The observations from the two main cases here are developed in collaboration with Ragnar Rosness, Kristine Størkersen, Jens Petter Johansen and Abdul Basit Mohammad.

Based on both my theoretical interests and the empirical data, the present discussion has value for some contexts and topics more so than others. Industrial safety is a matter of avoiding small accidents and incidents as well as larger events. This chapter is mainly, but not exclusively, about situations in which the work itself is critical for safety and reliability. The quality of the work of ship captains, infrastructure technicians and control room operators is in itself relevant for safety and/or reliability. A typical setting for this discussion is an information-dense control setting where there is some catastrophic potential, the bridge of a ship, a cockpit of a plane, or a control room. In other work situations, safety may be more loosely associated with the quality of the work itself. A related delimitation of this discussion is that we are mostly concerned with safety and reliability with regards to major accidents and incidents of a more systemic nature. The questions inspiring this book concern how one can train employees to be safer and implement policies to improve safety. In this respect there is a difference between simple injuries (a worker falling down the stairs or bumping his head) and more systemic and complex system breakdowns. This chapter, and the findings reported here, is skewed towards the latter type of incidents. Lastly, there is an implicit assumption of good intentions in my argument. In the cases I have studied, both the management and workers have great interest in prioritizing safety and some leverage to achieve it. Sometimes that is just not the case.[2]

7.2 Briefly on the Theoretical Background

Suchman's (1987) book *Plans and situated action* is a cornerstone in ethnographically-oriented studies of work, and a central reference point for my discussions of the relationships between situated work and representations of it. Her book and related theory based on detailed studies of work tend to highlight the uniqueness of situations, and thus provide a necessary counterweight to organizational theory and management perspectives. While studies of "situated practice" can be seen as an insistence that procedures and plans do not represent action, this is only half the story. They should also be considered as a call to see the pragmatic role of these representations, the tools they constitute, *in situations*. The way practice is intertwined with material and symbolic artefacts in situated work represents another part of the theoretical background for this chapter.[3] This is inspired

---

[2]The underlying causes for the South Korean Sewol ferry accident show how several actors seem to have a weak interest in safety (Kim et al. 2016). At the workshop (organized by FonCSI in November 2015 and highlight of the project that led to this book, editors' note) Jonathan Molyneux presented a rather grim picture with regards to the available resources for addressing safety in the global mining industry.

[3]See also Gherardi in this volume on the relationship between situated practice and safety.

by several sociotechnical approaches to situated practice.[4] One such, is Hutchins'
(1995; Hutchins & Klausen 1996) discussions of "distributed cognition", a strand of
theory that stresses the relations between technologies and representation and
thought, to the extent that the primary object of study is the distributed system.
Understanding the always unique nature of situated action also fits well with recent
safety theoretical frameworks like Resilience Engineering, which stresses the
importance of ever-present variability, and how one performs work in situated
contexts to handle it.[5] In this literature Suchman's plans and situated actions have
their counterpart in "work as imagined and work as done" (Dekker 2006; see also
Hollnagel 2015; Nathanael and Marmaras 2006; Haavik 2014).

A key trend in organizational life today is the increased focus on accountability
and auditability. In the "audit society" (Power 1997) control, including over risk, is
sought through standardization, measurement and counting (Power 2007; Hohnen
and Hasle 2011; Almklov and Antonsen 2010, 2014). If possible, work is broken
into manageable entities to be controlled by bureaucratic methods (such as audits or
"management by objectives") or market-based means. Tasks are delimited and
decontextualized as much as possible in order for them to fit with audit schemes.
The resulting paper trails can be used to make workers and managers "accountable"
for safety. Of course, some things are easier to standardize and control in this way.
More complex and situationally contingent work is hard to standardize (Almklov
and Antonsen 2014), and much of what we regard as professional competence is
left out. Moreover, the whole doctrine of accountability tends to skew our attention
towards anticipating known risk, rather than being open for the unknown (see
Wildavsky 1988). The first cases I will describe are examples of how safety, under
the   global   developments   towards   standardization,   accountability   and
self-regulation, has become an organizational discourse where generic models
dominate over insights into the contextual peculiarities of different industries and
work contexts.

---

[4]A somewhat idiosyncratic selection of mine would include studies from science and technology
studies (e.g. Latour, 1999; and my own take in Almklov, 2008), distributed cognition (Hutchins,
1995; Hutchins & Klausen, 1996) anthropology of technology (Ingold, 2000) and sociomaterial
theory (Orlikowski & Scott, 2008). All of these have different, but some sort of *relational,*
conceptions of representation and technology.

[5]Similar thoughts are also found in the literature on High Reliability Organizations (LaPorte and
Consolini 1991; Weick and Sutcliffe 2015). It is most explicitly argued in Resilience Engineering
(Hollnagel et al. 2006). Also within the field of ergonomics the distinctions and relationships
between representations of work and work as performed has been theorized (e.g. Guérin et al.
2007).

7.3 First Example: Compartmentalization of Safety in Shipping and Railroads

Sometimes analytical ideas can be located in time and space. This comes from Bergen, Norway in 2012.

In 2012, Kristine Størkersen and I conducted interviews with the Norwegian Association of Cargo Freighters as a part of a project on regulation and safety culture in the transport sectors. This visit followed several interviews onboard transport ships and passenger vessels. Compared to the mighty Norwegian Shipowners' Organization in Norway, that represents the international shipping industry, this interest organization is small and modest. Our interviews concerned how regulation of shipping influenced safety culture in shipping. In particular, we ended up discussing the ISM code, the international system dominating the management of safety on ships around the globe. The ISM code is an international standard requiring every ship to have a safety management system. It is built around principles of self-regulation, but it also places several demands on these systems. The ISM code is developed by the International Maritime Organization (IMO). The organization we visited represented several small (and a few larger) ship owners, many of them family businesses with one or two ships, and the ships themselves varied in size and technical complexity. Our interviews in this organization centered on the tension between the global standards, represented by the ISM code and the practical reality onboard some of these ships. Throughout the industry, the standard was seen as demanding way too much in terms of paper work and of being of little practical use, being hard to adapt to the practical reality. One interviewee exemplified this for us by describing how some sand boats operated in Norwegian fjords, basically sailing back and forth with sand or gravel from a quarry with a crew of two to three. And yet, he sighed, these boats are essentially under the same legislation as an oil tanker, so the inspector "should have some sense of reality!" Most ships needed consultants to help them develop a safety management system, and the systems they developed were typically too generic and too complicated to be of practical use. The inspections by national authorities (through classification societies or directly by the regulator) also focused on compliance with the ISM code and that the paperwork was in order, i.e. that they had a compliant safety management system (SMS). Thus, the discourse of safety drifted towards a system of auditable items, satisfying the ISM standards, and then complying with them.

Several of the employees at the association had worked on ships, and they cooperated closely with captains and shipowners. They, like the seamen we had interviewed earlier in the project, lamented the distance between the safety management systems implemented to control safety and the practical realities onboard the ships. Most systems were primarily paperwork, something that they were required to comply with, with little practical relevance for the operational safety.

Moreover, the shipowners and captains were caught in a principal-agent[6] relationship with the consultants. Consultants, moving around from ship to ship, may be fine with a generic and large safety management system, while the seamen that are supposed to use it and pay for it would prefer a simpler system, and one more adapted to their operational context. The ship owners' interest organization, recognizing this, had developed their own consultancy service to help their member shipping companies develop less complicated systems tailored to their needs (while still fulfilling the minimal demands of the ISM code). What we observed, and which became so clear for us during interviews with these "translators", was the compartmentalization of safety the ISM code and the safety management systems had led to. The well-intentioned efforts towards improving safety demanded a system of governance that was so complicated that the practitioners were unable to handle it, and had to resort to consultants processing the paperwork for them. The knowledge of individual ships, on how to operate them and the risks that this implies, became subordinate to a formal generic system. Moreover, handling the interface between the system of governance and practice depended on another form of expertise.

This was also apparent in our interviews with captains, ship owners and crews. The demand for documentation and reports took attention away from the key tasks of the seamen, and was particularly problematic in small businesses. Moreover, the paperwork[7] was not aligned with their professional practices as seamen. Rules and procedures were too specific and too little adapted to their work context and skills to be useful. The SMS didn't help them in the most important parts of their work. A sailor on an anchor handler, a strong tugboat working for the petroleum industry, described the lack of relevance of the SMS to me during a break between activities on deck in an operation at an offshore oilfield. Being on deck on an anchor handler is truly hazardous work involving heavy machinery, chains, wires and winches. There wasn't much paperwork with this work, he told me, but as soon as the ship is anchored in the harbor and he wants to do some painting there are all sorts of forms to fill out. The procedures were most relevant for the least dangerous work, and then they didn't make much of a difference anyway according to him. Another informant on a high speed passenger craft noted how the SMS describes how to mark out routes in a way that didn't consider weather and current, commenting:

> experienced navigators want to – and do – choose a course according to wind and current.[8]

In his organization, operating a fleet of High Speed Passenger crafts, they had answered the demand for reporting and a solid safety management system by employing safety professionals onshore. Many of these professionals had experience from other industries and a more generic and systems-oriented approach to safety. Though there are nuances to this image, we recorded numerous examples in

---

[6]See Eisenhardt (1989) for an introduction to Agency Theory.

[7]See Knudsen (2009) for a discussion of the relationship between paperwork and seamanship.

[8]This example is also discussed in Størkersen et al. (2016).

this and other projects of how the safety management system was regarded having little relevance for the core activities onboard the ships. In both these examples, one may assume that this lack of relevance has to do with how the professional competence of the users, of captains and deck hands, is about navigating within dynamic and situationally contingent situations. A generic "recipe" on how to behave on deck during an evolving anchor handling operation will just not capture the essence of this dynamic and situationally contingent work.

In the resulting paper (Almklov et al. 2014) we also included Ragnar Rosness' historical account of the Norwegian Railways. There too, the development towards a more "professional" approach to safety, or "Health Safety and Environment", led to a discursive dominance of what one may call "theoretical" or generic approaches to safety. This can be traced as a historical development through several organizational changes and reorganizations where the railroads' traditional "Safety Office", specializing on how to build and operate the train system safely, gradually became subordinated to an HSE department consisting of safety experts from other industries, specializing in more generic models of safety. The once so powerful safety office moved downwards in the hierarchy in the organizational model. Their perspectives on how to make the railroad system safe became less important, and less significant in the organizational discourses. Several mechanisms contributed to this. For example, since investigations after accidents were typically based on generic models of safety, inspired by other industries, the need for more systematic and accountability-based approaches to safety tended to be the obvious measures to implement afterwards. The railroad-specific safety knowledge was still there, but its proponents were less powerful, and consequently resources were directed towards other forms of safety. In both cases we observe a weakening of the practitioners perspectives in safety management. These are some possible downsides of strengthening safety as a separate discipline. If the object of interest is safety, it is easy to ignore or lose track of the peculiarity of the operational contexts.

7.4 Second Example: Anticipatory Work in Space Operations

The control room operating a research module at the international space station (ISS) is a fascinating study object for research on reliability and resilience (see Fig. 7.1). However, going beyond the control room itself, and including details of the surrounding organizational processes, preparation, planning and training, is even more interesting.

These other activities also, our informants repeatedly reminded us, makes up more than 90% of their work. When you work with advanced and costly space operations, reliability and resilience is at the very core of the work activities. My colleagues and I followed the work of a team of research engineers conducting a

Fig. 7.1 Two research engineers watch as an astronaut at the ISS injects water into experiment containers (each with an individual seed) according to a detailed procedure they have developed and verified in advance

biological experiment on the ISS, and we studied their extreme focus on anticipating and mitigating possible problems in advance.

The control center N-USOC[9] is part of a distributed network of small control rooms operating individual equipment onboard the ISS. This control room's most important payload is a microgravity research laboratory used for biological experiments on plants. The research engineers at N-USOC can be seen as a form of lab technicians, helping researchers transform ideas into workable experiments, testing and verifying equipment and procedures before the seeds are sent to the IS. Then they monitor the experiment as it is conducted. Due to the high cost, low accessibility and low tolerance for risk,[10] space operations is an interesting case for studying reliability and resilience. Every trivial detail that could possibly cause a problem is subject to intense scrutiny. In the paper "What can possibly go wrong?" (Johansen et al. 2015) we identify and discuss "anticipatory work": practices constituted of an entanglement of cognitive, social and technical elements involved in anticipating and proactively mitigating everything that might go wrong.[11] The nature of anticipatory work changes between the planning and the operational phases of an experiment.

---

[9]The Norwegian User Support and Control Centre.

[10]E.g. any risk of the experiment polluting the atmosphere of the ISS or harming the astronauts is unacceptable.

[11]Recently similar types of sociotechnical work have been labeled "anticipation work" within STS. See Steinhart and Jackson (2015) and Clarke (2016).

The case revolves around an incident where the control room operators have to solve a telemetry error. The data from the lab module fails to reach the control room. This threatens to ruin a multi-million dollar experiment that has been planned and prepared for seven years. We followed the resolution of the problem. But, importantly, we had also studied the anticipatory work that this troubleshooting relied upon. In this preparatory stage, every anomaly that has happened in previous experiments is analyzed and mitigated in advance, either by technological changes, by changing computer scripts, or writing "just in case" scripts, by developing procedures or protocols. An informant explains:

> First of all it is things that have happened before and we know can happen again. After that we just sit and think 'what if that happens, even though it looks impossible?', so we start to think very negatively, that works well, and we write what-if scenarios.

Throughout the planning phase possible problems that could occur were identified and subject to collective reflection. They were documented and possible solutions were developed. The telemetry error they experienced had been experienced before. They did not know exactly what caused it, and could not fix it permanently, but they had developed several procedures that might fix it.

Problem resolution in the operational phase definitely resembled the typical story in safety journals on control room operations. There was a process of confusion and ad hoc-sensemaking as they tried to understand the problem. The process also demanded some creative thinking. However, the cognitive and social process in the operational phase is intrinsically connected to the anticipatory work conducted in the planning phase. The critical difference being that the solutions developed in the calm of the preparatory phase had to be situated in the temporal flow and situational contingencies of the real-time phase. The first solution was to send a pre-programmed work-around script to the unit. This is minimally invasive and something the N-USOC can do without involving entities from the NASA/ESA network, which they did after that they had diagnosed the problem. However, this work around was unsuccessful. The next procedure was to restart a computer on the ISS handling the telemetry data. To do this, they would have to coordinate with other entities at ESA and NASA. Even though these preplanned fixes had been worked out in detail, their plans could not take into account parallel activities at the ISS. Thus a key task for the operators is to use their understanding of the interaction effects with other operations and systems and find a way to execute this reboot in an acceptable manner. Unfortunately, another greenhouse experiment was active with ongoing astronaut activities that continued for some time, and N-USOC couldn't restart the computer before that had been completed, since the other team's equipment was connected to it as well.

The temporal dimension complicates the matter further in several ways:

1. their own experiment cannot continue without telemetry for much longer, so it is urgent to get it fixed,

2. communication with the ISS only works in irregular, but pre-identified intervals,[12]
3. and of course, they are unable to control the speed of the other experiment blocking their reboot.

Thus, they need to look for upcoming time-slots to perform their shut down, as soon as the other experiment is done. This is something they have not pre-planned, but their pre-planning of solutions is crucial for their resolutions, as it provides them with pieces of the temporal puzzle. They improvise with plans, and this improvisation is mainly about situating the plans in a temporal flow. Moreover, in their interaction with important stakeholders in the ESA and NASA hierarchy being able to refer to pre-planned interventions fast-tracks their go-ahead for the restart.

By focusing not only on the control room activities as the experiment unfolded, which we recorded on video and analyzed in detail, but also on the organizational context and extensive preparations, we made two observations with implications for the governance of safety. We demonstrate in some detail how the engineers try to anticipate upcoming contingencies and how they produce solutions to these— technological fixes, procedures, checklists, etc. and how these become parts of a sociotechnical body of knowledge. The procedures and fixes are indivisible parts of their "distributed cognition" (Hutchins and Klausen 1996). The actions of the control room operators are located in a situation where procedures, protocols, checklists, computer scripts etc. are an intrinsic part. The debates in safety research on the extent to which rules and procedures can or should control practice, must be nuanced with a discussion of whether these are an integrated part of practice or not. In this case they are, and procedures and practice are entwined, but in other cases procedures mainly serve management purposes. We saw how this seemed to be the case in shipping, and we have also seen similar developments in petroleum (see for example Antonsen et al. 2008, 2012). Due to the dominating logic of accountability, control by standardization and compartmentalization of HSE, the representations of work are (often) too decontextualized to be of much use in situated work contexts.

A second observation with relevance for this book is the implications of the different temporalities of the planning phase and the operations phase. In the operation phase of the experiment, plants have been watered and are growing, so time is running unstoppably. The operators continuously try to stay ahead of unfolding events and coordinate with parallel activities. They cannot turn back, and must continuously improvise to implement even the best-laid plans. This work clearly fits the typical narrative in resilience engineering. It is about handling variability and navigating uncertainty not only to avoid errors. In the planning phase, however, the anticipatory work is indeed characterized by an intense focus on "what can possibly go wrong". The tolerance for errors is very low (due to the

---

[12]The communication coverage is displayed on a timeline that is usually displayed on one of the control-room screen to allow the operator to be aware of upcoming communication shadows before initiating activities or data transfers.

cost and low accessibility of the space station) so extensive work is undertaken for mitigating every possible contingency in advance. The differences in terms of temporalities of these two phases, and the practices that make them safe, require different strategies of management and training.

7.5 Discussion: Some Propositions

In sum I have put forward some ideas based on studies of situated work in critical settings. While I have exemplified these ideas with observations from shipping, railways and a control room, they are not solely based on these settings.

The mode of control in modern organizations, centered around standards, accountability and a decontextualized view on practice, could render important aspects of practice less visible, and discursively weaker. The drift towards more generic and accountability-centered approaches to safety can make procedures increasingly decontextualized, and decoupled from practice. However, some of the aspects of work that are "invisible" in this discourse of work, such as adapting to the variability of concrete situations, are important for resilience and reliability. Thus, important parts of what makes work safe are often not regulated or supported in the installed safety management systems, due to their situation-specific nature. Increasing the granularity of the existing systems, regulating work in even more detail, is not likely to improve that.

It is important to note that procedures, rules and checklists can be an integrated part of a community of practice, a *resource* for improvisation, a means of remembering shared knowledge, and an inextricable part of the "distributed" knowledge of the workers. Other times, they primarily serve purposes of accountability and external control. Discussions of rules and procedures (see e.g. Hale and Borys 2012) and how they contribute to safe practice should distinguish between these functions. It is not a matter of rules versus improvisation, but of how rules and procedures may support or hamper situational improvisation. For managers, a consequence of this insight should be to resist, or at least reflect critically on, the temptation to integrate procedures that work in one setting, within one community of practice, with the company's more generalized safety management systems. Secondly, managers should seek to understand the situationally adaptive work that is necessary in critical work processes, recognize that this work might be impossible to standardize and enroll in organizational systems of control. However, it still needs to be supervised.

The temporality of the work situation is an important factor in understanding the relationship between representations of work and situated practice. In some types of work, such as the work of control room operators described here, the petroleum processing plant operators described by Kongsvik et al. (2015) or infrastructure technicians described in Almklov and Antonsen (2014), creatively situating planned activities in a temporally unfolding situation is a core task. In all these settings, the workers deal with unique situational contingencies. This fits poorly in rationalistic models of work and can be invisible in formal descriptions. Generally, representations of work tend to be detached from the evolving temporal trajectories of work as performed. A process that goes on and on, like the seedling growing in a greenhouse on the space station or a process plant running continuously, has a temporal trajectory that must be considered. There are temporal constraints on decisions and work execution. For example: simultaneous activities that might influence your activities or system, people getting tired over time, shifts ending, there is a difference between doing the same task the first time from the second time, etc. In operational work, managing such temporal trajectories and handling temporal variability is crucial, both for getting work done and getting it done safely.

One caveat, however, is that the accounts and theorizing about improvisation and the handling of variability in such situations should not be uncritically employed in work in situations with other temporal characteristics. Sometimes, like in the planning phase of the space experiment, one has the time and takes the time to plan and re-plan to avoid everything that could possibly go wrong. And sometimes a standardized description of a task is almost all you need. Arguably, many of the insights generated in recent years in safety science, e.g. in Resilience Engineering, on the importance on managing variability, are mostly relevant in *operational* settings, within an operational temporality and with a certain amount of situational variability. Thus, for managers and workers seeking to improve safety, recognizing the difference in temporality of different settings is an important step in choosing strategies for safety management for each situation.[13] One should not be trying to model one in the image of the other.

Many organizational discourses and systems implemented to improve safety are centered on standardized tasks and measurable goals and they fail to capture important aspects of what makes work safe. This book is about professionalization of safety, on how to improve safety even further in industrial settings. A key argument of this chapter in this respect is that the systems, procedures, rules, checklists and reports supporting work in operational settings must be developed with a keen eye on the situational improvisation and adaptation that is often important in such work, not only for its efficient execution but also for its safety.

References

Almklov, P. G. (2008). Standardized Data and Singular Situations. *Social Studies of Science, 38*(6), 873–897.

Almklov, P. G., & Antonsen, S. (2010). The commoditization of societal safety. *Journal of contingencies and crisis management, 18*(3), 132–144.

---

[13]Journé and Raulet-Crouset (2006) interestingly discuss how to manage situations, or situated work, including the role of temporal structures. Also Hayes (2012) suggests ways to balance pre-planned operational envelopes with ways of managing safety in evolving situations. More generally Grøtan (2017) presents interesting models for the management of resilient practice.

Almklov, P. G., Antonsen, S. 2014. Making work invisible: new public management and operational work in critical infrastructure sectors. *Public Administration, 92*(2), 477–92.

Almklov, P. G., Rosness, R., & Størkersen, K. (2014). When safety science meets the practitioners: Does safety science contribute to marginalization of practical knowledge?. *Safety science, 67*, 25–36.

Antonsen, S., Almklov, P., & Fenstad, J. (2008). Reducing the gap between procedures and practice–lessons from a successful safety intervention. *Safety Science Monitor, 12*(1), 1–16.

Antonsen, S., Skarholt, K., & Ringstad, A. J. (2012). The role of standardization in safety management–A case study of a major oil & gas company. *Safety science, 50*(10), 2001–2009.

Barley, S. R., & Kunda, G. (2001). Bringing Work Back In. *Organization Science, 12*(1), 76–95.

Clarke, A. E. (2016). Anticipation Work: Abduction, Simplification, Hope. *Boundary Objects and Beyond: Working with Leigh Star*, 85.

Dekker, S. (2006). Resilience engineering: Chronicling the emergence of confused consensus. In E. Hollnagel, D. D. Woods & N. Leveson (Eds.), *Resilience engineering: Concepts and precepts*. Hampshire: Ashgate.

Eisenhardt, K. M. (1989). Agency theory: An assessment and review. *Academy of management review, 14*(1), 57–74.

Grøtan, T.O. (2017). Training for operational resilience capabilities (TORC): Summary of concept and experiences. Trondheim: SINTEF Report A28088.

Guérin, F., Laville, A., Daniellou, F., Duraffourg, J., & Kerguelen, A. (2007). *Understanding and transforming work. The practice of ergonomics*. Lyon: Anact Network Edition.

Haavik, T. K. 2014. Sensework. *Computer Supported Cooperative Work (CSCW), 23*(3), 269–298.

Hale, A., Borys, D. 2012. Working to rule, or working safely? Part 1: A state of the art review. *Safety Science*.

Hayes, J. 2012. Use of safety barriers in operational safety decision making. *Safety Science. 50*(3), 424–432.

Hohnen, P., & Hasle, P. (2011). Making work environment auditable–A 'critical case' study of certified occupational health and safety management systems in Denmark. *Safety Science, 49*(7), 1022–1029.

Hollnagel, E. (2015). Why is work-as-imagined different from work-as-done. In R. L. Wears, E. Hollnagel, & J. Braithwaite (Eds.), *Resilient Health Care: The resilience of everyday clinical work* (pp. 249–264). *Farnham, UK*: Ashgate.

Hollnagel, E., Woods. D.D., Leveson. N. (2006). *Resilience engineering: concepts and precepts*. Gower Publishing Company.

Hutchins, E. 1995. *Cognition in the wild*. Cambridge: MIT Press.

Hutchins, E., Klausen, T. 1996. Distributed cognition in an airline cockpit. In: Engeström Y, Middleton D (Eds), *Cognition and communication at work* (pp. 15–34). Cambridge: Cambridge University Press.

Ingold, T. (2000). *The perception of the environment: essays on livelihood, dwelling and skill*. Psychology Press.

Johansen, J. P., Almklov, P. G., & Mohammad, A. B. (2015). What can possibly go wrong? Anticipatory work in space operations. *Cognition, Technology & Work*, 1–18.

Journé, B., & Raulet-Croset, N. (2006). The concept of situation: a key concept in the studying of strategizing and organizing practices in a context of risk. Paper presented at EGOS 2006.

Kim, H., Haugen, S., & Utne, I. B. (2016). Assessment of accident theories for major accidents focusing on the MV SEWOL disaster: similarities, differences, and discussion for a combined approach. *Safety science, 82*, 410–420.

Knudsen, F. (2009). Paperwork at the service of safety? Workers' reluctance against written procedures exemplified by the concept of 'seamanship'. *Safety science, 47*(2), 295–303.

Kongsvik, T., Almklov, P., Haavik, T., Haugen, S., Vinnem, J. E., & Schiefloe, P. M. (2015). Decisions and decision support for major accident prevention in the process industries. *Journal of Loss Prevention in the Process Industries, 35*, 85–94.

LaPorte, T.R., Consolini, P.M. 1991. Working in practice but not in theory: theoretical challenges of "high-reliability organizations". *Journal of Public Administration Research and Theory, 1*(1), 19–48.

Latour, B. 1999. *Pandora's hope: essays on the reality of science studies*. Cambridge: Harvard University Press.

Nathanael, D., & Marmaras, N. (2006). The interplay between work practices and prescription: a key issue for organizational resilience. In *Proc. 2nd Resilience Eng. Symp* (pp. 229–237).

Orlikowski, W. J., & Scott, S. V. (2008). Sociomateriality: challenging the separation of technology, work and organization. *The academy of management annals, 2*(1), 433–474.

Power, M. (1997). *The audit society: Rituals of verification*. Oxford: Oxford University Press.

Power, M. (2007). *Organized uncertainty: designing a world of risk management*. Oxford: Oxford University Press.

Steinhardt, S. B., & Jackson, S. J. (2015). Anticipation work: Cultivating vision in collective practice. In *Proceedings of the 18th ACM Conference on Computer Supported Cooperative Work & Social Computing* (pp. 443–453). ACM.

Størkersen, K. V., Antonsen, S., & Kongsvik, T. (2016). One size fits all? Safety management regulation of ship accidents and personal injuries. *Journal of Risk Research*, 1–19.

Suchman, L. 1987. Plans and situated actions. New York: Cambridge University Press.

Suchman, L. A. (1995). Making work visible. *Communications of the ACM, 38*(9), 56–64.

Weick, K. E., & Sutcliffe, K. M. (2015). *Managing the unexpected: Sustained performance in a complex world*. [Third edition] John Wiley & Sons.

Wildavsky, A. B. (1988). *Searching for safety*. Transaction publishers.

Chapter 8
Stories and Standards: The Impact of Professional Social Practices on Safety Decision Making

Jan Hayes

Abstract Organisational influences on safety outcomes are the subject of much attention in both academia and industry with a focus on how workplace factors and company systems, both formal and informal, influence workers. Many individuals who make important decisions for safety are not simply employees of a particular firm, but also members of a profession. This second social identity is little studied or acknowledged and yet is it critical for safety. This chapter addresses two key social practices that influence safety outcomes. The first is professional learning for disaster prevention. Research has shown that much professional learning is profoundly social including sharing stories and using stories directly as an input to key decisions. Another critical professional activity is development of standards. Standards are seen as authoritative sources and so 'called up' in legislation and yet the processes by which they are developed are opaque to those outside the small group of professionals involved. Again, this important social practice of groups of professionals remains little studied. Professional social practices such as these are worthy of much more attention from both academia and industry.

Keywords Professional · Safety imagination · Technical standards · Storytelling

8.1 Introduction

Many individuals working in industries with the potential for disaster such as offshore oil and gas, chemicals, aviation and nuclear power make decisions that can ultimately have a major influence on safety outcomes. Making good safety decisions is significantly reliant on technical skills and knowledge and, for any complex

J. Hayes (✉)
RMIT University, Melbourne, Australia
e-mail: jan.hayes2@rmit.edu.au

© The Author(s) 2018
C. Bieder et al. (eds.), *Beyond Safety Training*, Safety Management,
https://doi.org/10.1007/978-3-319-65527-7_8

technology, safe operation requires expert knowledge in a wide range of specific fields to be brought together. Safe operation of commercial aircraft, for example, requires expertise in various engineering disciplines to design the aircraft in addition to expert pilots, aircraft maintainers and air traffic controllers etc. to ensure safety once any aircraft is in service. Each of these specialisations can be seen as a profession, built around a particular body of expert knowledge that is strongly linked to one or more aspects of safety performance.

When things go wrong, investigations into the causes of major disasters rarely reveal new technical knowledge but rather show that existing knowledge was not applied. The reasons for this are invariably social (Weick and Sutcliffe 2001). Preventing accidents in all cases requires that those who are making safety related decisions are aware of the potential consequences of their actions and so make choices in that light. This attitude to work, having a safety imagination (Pidgeon and O'Leary 2000), is the opposite of complacency and is reflected in another quality of the professions—a sense of being worthy of public trust (Middlehurst and Kennie 1997).

Professionals learn in formal settings as they gain formal qualifications and attend training courses throughout their career. Critically, key professional knowledge also originates from on-the-job working with other members of their professional group. Some technical tips are learned this way but learning to have a safety imagination in particular takes place in a group integrated with daily activities in a 'community of practice' (Wenger 1998).

In contrast, key technical knowledge is often recorded and communicated in a very different form—that of technical standards. Such standards are important for professional learning, although those with a good safety imagination understand that compliance with standards alone is not an adequate safety strategy. Professional groups also have a key role in technical safety assurance as custodians of the content of key technical standards. Whilst standards often have pseudo-regulatory status, processes of standard formation have been little studied and are often opaque. Since some influential standards are produced by organisations that have a primary function as industry lobby groups, this area also deserves further study and critical attention.

This chapter explores the links between professionals and learning for safety. In organisations under increasing cost pressure, time for professional activities is often seen as not a core part of company activities and yet organisations rely critically on professional judgement to ensure that operations continue safely. Many organisations fail to recognise this and instead see safety as grounded in company systems, rather than professional practice.

Drawing on previously published research across a variety of industrial domains, key points are highlighted regarding how professionals make safety-related decisions, the role of experts and professional societies in standard formation and the implications for organisations and safety decision-making.

8.2 Expertise, Professionals and Learning in the Context of Disaster Prevention

The starting point for this chapter is that for lessons of past accidents to be learned, i.e. taken into account in future decision-making, experts must maintain a 'safety imagination'(Pidgeon and O'Leary 2000). A lack of safety imagination is linked to a psychological rigidity that restricts decision makers in their ability to link their work to the possible consequences. The question is therefore how a safety imagination can best be fostered. Researchers in the field of naturalistic decision-making (e.g. Klein 1998) have identified stories as an effective knowledge source for decision-making in critical contexts, because they are a powerful tool in pattern matching and mental simulation. Stories convert experiences into memorable, meaningful lessons (Klein 1998; Polkinghorne 1988). As Schank (1990) puts it,

> We need to tell someone else a story that describes our experience, because the process of creating the story also creates the memory structure that will contain the gist of the story for the rest of our lives.

Storytelling is fundamentally a social practice. Table 8.1 contrasts social learning with more traditional learning approaches linked to formal training. Social learning, which includes attitudes and behaviours as well as facts, is ongoing, action-oriented and collaborative.

Experts keep their knowledge up to date by ongoing social learning from professional peers (Dreyfus and Dreyfus 1986). In an occupational sense, expertise is one of a number of qualities displayed by a professional. Others include exercise of trust as the basis for professional relationships, adherence to defined professional ethics and independence (Middlehurst and Kennie 1997). Other authors (Friedson 2001; Sullivan 2005) expand on these ideas to describe the strong sense of responsibility held by professionals for the public good. Professions have strict entry standards in the form of long training in both theoretical and practical considerations and often licensing arrangements. This training and induction into the

Table 8.1 Models of learning

Traditional	Social
Individual	Group/organisation
Isolated from workplace	Integrated with daily activity
About facts	About attitudes and behaviours, as well as facts
Collaboration is cheating	Working together is key
Teacher/student	Collaboration
Absolute	Context-dependent
Has defined beginning and end	Ongoing
Knowledge-oriented	Action-oriented

Adapted from Wenger (1998)

culture of the profession engenders members of this exclusive group with loyalty to their peers, rather than to their employers.

These links between safety imagination, learning and professionalism have an impact on safety outcomes but go largely unrecognised in organisations that see all their employees as simply that, rather than members of a profession. This has important safety implications as described below.

8.3 Professionals at Work

Most modern organisations are highly bureaucratised. Management is the dominant profession and the most senior managers, i.e. those at the top of the organisational hierarchy, set the goals of the organisation and the methods by which those goals will be achieved. Top managers are very powerful and the role of other members of the organisation is simply to implement strategies that are determined at the top—essentially, to follow instructions.

For organisations that operate hazardous technologies and have a high level of safety performance, this is only a partial view of the way power is distributed. In such organisations, professional groups other than managers also have significant power and authority when it comes to safety decision-making. Think of an airline pilot in the cockpit. A professional who holds this role operates within company systems, but also holds the ultimate decision-making authority for the safety of the aircraft. The most senior professionals in many organisations hold similar levels of power within their own particular domain. Expert design engineers, for example, hold significant authority and can exert it informally or by formal systems such as sign off of drawings and specifications that is required for projects to proceed. The question then arises as to how professional judgements are formed.

In our research, we have found senior operational staff sharing stories of their experiences in order to support professional judgements in three key ways (Hayes 2013a; Hayes and Maslen 2015). The first was directly linked to the concept of safety imagination. These are stories of past events where the moral of the story is the uncertainty of the technology and the need for vigilance to ensure that workers and the public are protected. We found people in fields as diverse as air traffic control, nuclear power station operations and chemical production telling similar tales to their colleagues. Less predictably, perhaps, was the fact that such stories did not always involve a major catastrophe (or a near miss). Some stories were simply about the significant unpredictability of the way the system behaved at some given point in time. After the event, such an experience is shared by telling and retelling. Weick calls such an event a cosmology episode,

> A cosmology episode occurs when people suddenly and deeply feel that the universe is no longer a rational, orderly place. (Weick 1993, p. 633)

Such incidents may be professionally life-changing for the person involved and of great interest to fellow professionals but possibly insignificant for non-experts.

Stories such as these are very popular within a specific community of practice. We have found that they are used by individuals to develop and reinforce their own safety imagination but also to foster such attitudes within less experienced members of the professional community. As we found one senior design engineer asking his much younger colleagues as he set out to check their calculations,

who have you killed today? (Hayes 2015b)

Story-based tests such as these were used by technical professionals when coming to a decision in the design office and also in an operating environment. Experts make decisions intuitively and one way to tap into their intuition is to imagine themselves in such situations as:

- walking into the plant following a proposed change with their young child in their arms,
- having to call someone's family to explain why they are injured if the planned work were to go ahead and then something went wrong,
- seeing their decision published in the media.

These methods sit alongside formal company systems such as risk assessment and influence outcomes, yet they are not acknowledged.

The third way in which we found professionals sharing stories was directly in the form of technical lessons learned about the system that they operate. Even these stories are not what organisations typically think of as incidents to be reported, recorded in a database and analysed for future learning. Rather, they are stories of small anomalies in system behaviour that are of particular interest to a specific group of people.

In all three cases it is clear that a story is much more than simply an accumulation of facts. Stories link human actions and events into an integrated composite (Polkinghorne 1988). As such, they have protagonists, a narrative and causal relationships. These features assist both senior operating professionals and design engineers to develop and maintain a safety imagination. More than that, sharing such stories is a profoundly social practice which gives them the professional courage to deliver bad news upwards, knowing that it may not be well received. As one design engineer told us,

[Senior management] will support the calls I've made. They are disappointed and I'm not always popular but they see the importance. (Hayes 2015a)

Senior operating professionals and expert discipline engineers in a design office may have a high degree of informal influence with senior management but this is often not visible in formal organisation charts. They have all chosen what Zabusky and Barley (1996) call a 'career of achievement'—where status and seniority is judged by others based on professional skills and knowledge. Whilst they are at the top of their chosen profession, they may only appear in the middle levels of the formal organisation chart. In contrast, those higher up the organisation chart with a greater level of seniority in strict hierarchical terms have chosen a 'career of advancement', but likely they have left their chosen profession behind and become primarily managers instead.

It might be challenging for organisations to acknowledge that their key technical staff have an allegiance to their profession which is at least partially independent of their allegiance to their employer. Nevertheless allowing time for, and engaging more directly with, professional activities is an untapped mechanism by which companies can influence safety outcomes.

8.4 The Role of Standards

Another key facet of technical professionals' practice is a high degree of respect for technical procedures and standards, along with an awareness of their limitations. Particularly in the field of engineering design, key information is often found in technical standards. Main and Frantz (1994) found that 87% of the design engineers they surveyed cited standards and codes as a key source of safety information (more than any other source). When it comes to excellence in safety outcomes, this raises important questions regarding how standards are used by engineers in making judgements that impact outcomes and also about the source of the information contained in standards.

Compliance with industry standards is a common legislative requirement. This makes sense when standards are 'experience carriers' (Hale et al. 2007). Compliance with rules (of which standards are one kind) is a well-recognised strategy for both constraining and supporting decision makers (Hale and Borys 2013a, b) but in the end if standards are not applied mindfully (Weick and Sutcliffe 2001) then safe outcomes cannot be assured.

A case in point was the failure of Enbridge's oil pipeline at Marshall, Michigan which caused the largest onshore oil spill in US history (Hayes and Hopkins 2014). The pipeline that failed was known to be severely cracked and yet engineers had put significant effort over years into demonstrating that the cracks did not meet the requirements laid down in the relevant standard that would have triggered repair. The standard specified requirements for pipeline cracks and different requirements for corrosion-related defects. Enbridge chose to treat the more than 50 inch long crack as a corrosion defect because it was initially caused by corrosion. This determination led them to use a different (less conservative) method to estimate the remaining strength of the cracked pipeline. They also embedded several other optimistic assumptions in their calculations. If they had treated the fault as a crack, calculations would have led to a different result and the line would have been excavated for physical inspection and likely repair. Instead, the line remained in service for an additional five years before it eventually failed with severe environmental consequences. Most relevant to our considerations here, the official investigation (NTSB 2012) came to the conclusion that company engineers had put significant effort into ensuring that the line complied with the standard, rather than considering whether the line was safe to leave in operation. These two issues are not identical.

In summary, whilst compliance with standards is important, if users lose sight of the reason for the requirements in the standard and come to see compliance as an end in itself, then safety is compromised. For complex systems, standards cannot cover every eventuality. Application of standards requires judgement and experience. As one design engineer told us,

> [For the younger engineers], experience is lacking and sometimes when they read the standard, they don't see the reason behind the requirements. They apply the standard just like a cook book. (Hayes 2015b)

This assumes of course that the requirements of the standard are adequate at least within the intended scope. Following any disaster, the relevant standards come under a great deal of scrutiny and yet processes of standard formation are little studied by safety researchers. In theory at least, standards are not driven by the interests of any individual operating company or by government. In practice, oil industry standards in the US have been criticized strongly by the Presidential Commission investigating the causes of the Deepwater Horizon incident. Their criticism relates specifically to the standards of the American Petroleum Institute (API). They report,

> As described by one representative, API-proposed safety standards have increasingly failed to reflect "best industry practices" and have instead expressed the "lowest common denominator"—in other words, a standard that almost all operators could readily achieve. Because, moreover, the Interior Department has in turn relied on API in developing its own regulatory safety standards, API's shortfalls have undermined the entire federal regulatory system. (National Commission on the BP Deepwater Horizon Oil Spill and Offshore Drilling 2011, Chap. 8)

At least in this case, standards produced by an industry lobby group have been found wanting. Other standards are produced by professional associations such as the American Society of Mechanical Engineers (ASME). A comparison of the way in which API and ASME see themselves and their role is pertinent. According to their web site,

> API is the only national trade association that represents all aspects of America's oil and natural gas industry.

In addition, the stated mission of the organisation is,

> to influence public policy in support of a strong, viable U.S. oil and natural gas industry.[1]

On the other hand ASME,

---

[1]http://www.americanpetroleuminstitute.com/GlobalItems/GlobalHeaderPages/About-API/API-Overview.

is a not-for-profit professional organization that enables collaboration, knowledge sharing and skill development across all engineering disciplines, while promoting the vital role of the engineer in society

and their mission is,

to serve diverse global communities by advancing, disseminating and applying engineering knowledge for improving the quality of life; and communicating the excitement of engineering.[2]

Clearly API is an industry lobby group whereas ASME is a professional society. The two organisations have significantly different interests which could be expected to be reflected in the documents that they produce and yet both API and ASME standards have pseudo-regulatory status in many jurisdictions. This blurring of distinction between professional societies and industry associations seems to be another way in which the value of professionals and professionalism is underestimated. Professional judgement is involved in both the production and use of standards.

8.5   Standards as a Social Construct

It may seem at first glance that these two influences on safety decision making are in conflict. Written material in a technical style in the form of a standard perhaps sits uncomfortably alongside oral traditions of story-telling as alternative sources of knowledge but in fact both standard production and use are also profoundly social activities.

Standards, along with other kinds of rules and procedures, have been subsumed by bureaucracy. They are perhaps mostly thought of in isolation, as dry but authoritative text used by a solitary individual. In fact, as touched on above, standards are written by groups of people who have specific interests and the details are subject to significant negotiation. Standards are not written in a narrative style. This makes the protagonists in the story of their creation invisible but they are no less influential in the ultimate outcome.

Use of standards is also a social process—they can be used and abused. As described above in the case of the Marshall pipeline failure the social norms at the company meant that the requirements of the standard were applied in a particular way that was ultimately catastrophic. Standards are not often used by individuals in isolation but rather by members of a professional group who interpret the requirements in their own idiosyncratic ways.

The clear implication of the social nature of standards is that professional values in general and safety imagination in particular have a direct impact on both the

---

[2]https://www.asme.org/about-asme/who-we-are/mission-vision-and-strategic-focus.

content and use of standards. Given the link between safety imagination and story-based learning, the theoretical gap between stories and standards is not as great as it might first appear.

8.6 Conclusions

This chapter has made the case that professionalism is more than just expertise in a technical sense. Professional attitudes (or lack of them) already impact safety outcomes but companies and researchers have paid this aspect of organisations very little attention. In the interests of safety, we should make this invisible work, visible.

There are several ways that organisations could do more to promote excellence in technical professionalism:

- Reward professional expertise by providing improved professional career paths. Many organisations claim to have a both technical and managerial promotion streams recognising the distinction between a 'career of achievement' versus a 'career of advancement' as discussed earlier, but few really deliver.
- Allow time for professional activities, including mentoring of younger professionals and involvement in standards development activities.
  'Professional development' has become a euphemism for more formal training courses. As described earlier, professional learning is a profoundly social activity. Many organisations are reluctant to allow time for such activities.
- Ensure that learning from incidents includes story-based learning, not just recording facts in a database.
  Storytelling and action-oriented social learning are key factors in identity construction. What we do is who we are. As Gautherau and Hollnagel (2005) describe it,

  becoming an expert is not only about learning new skills ... it is also about constructing an identity of a master practitioner.

- Give senior technical professionals more formal access to senior levels of management and reward bad news.
  In this chapter, we have focused primarily on decisions made by professionals that they then have the power to enact. Another key role of technical professionals is to advise more senior managers. Accident investigations have often shown that the failures that ultimately led to disaster were known about by some people but that the message was not transmitted upwards to those with the necessary power to intervene.

When organisational actors who should be professional fail to act in an appropriate way, the consequences can be literally disastrous. In the lead up to the

Montara blowout, drilling engineers who should have been providing professional oversight instead took a managerial approach to their role and left the technical details to the offshore crew (Hayes 2012, 2013b). This discussion is not meant to portray technical professionals as "white knights" and managers as "black knights". The interaction between these different professionals is more subtle than that. Finding the right balance between conflicting organisational goals—costs, schedule, safety—is difficult only because these are *all* legitimate concerns. Organisations can cope with some degree of cost overrun or production loss, and so, to some extent, these issues can be managed by trial and error, but when it comes to public safety, the challenge is to get the decision right every time. This requires the imagination to see what might go wrong and the foresight to see how it might be avoided without becoming so conservative that nothing is achieved.

The practical implications of the social side of standard formation are more difficult to comment on in a generalised sense because of the lack of research in this area. Standards tend to be seen as authoritative sources and in many ways they are but each one is also the product of a workplace. Given that we know how much the social and organisational aspects of work influence outcomes in other settings it is essential that processes of standard formation in particular receive further research attention.

Acknowledgements Aspects of this work were funded by the Energy Pipelines CRC, supported through the Australian Government's Cooperative Research Centres Program. The cash and in-kind support from the APGA RSC is gratefully acknowledged.

References

Dreyfus, H. L., & Dreyfus, S. E. (1986). *Mind over Machine*. New York: The Free Press.

Friedson, E. (2001). *Professionalism: The Third Logic*. Chicago: University of Chicago Press.

Gautherau, V., & Hollnagel, E. (2005). Planning, Control and Adaptation: A Case Study. *European Management Journal, 23*(1), 118–131.

Hale, A., & Borys, D. (2013a). Working to rule, or working safely? Part 1: A state of the art review. *Safety Science, 55*, 207–221.

Hale, A., & Borys, D. (2013b). Working to rule, or working safely? Part 2: The management of safety rules and procedures. *Safety Science, 55*, 222–231.

Hale, A., Kirwan, B., & Kjellén, U. (2007). Safe by design: where are we now? *Safety Science, 45*(1–2), 305–327.

Hayes, J. (2012). Operator Competence and Capacity - Lessons from the Montara Blowout. *Safety Science, 50*(3), 563–574.

Hayes, J. (2013a). *Operational Decision-making in High-hazard Organizations: Drawing a Line in the Sand*. Farnham: Ashgate.

Hayes, J. (2013b). The role of professionals in managing technological hazards: the Montara blowout. In S. Lockie, D. A. Sonnenfeld, & D. R. Fisher (Eds.), *Routledge International Handbook of Social and Environmental Change*. London: Routledge.

Hayes, J. (2015a). Investigating design office dynamics that support safe design. *Safety Science, 78*, 25–34.

Hayes, J. (2015b). Taking responsibility for public safety: How engineers seek to minimise disaster incubation in design of hazardous facilities. *Safety Science, 77*, 48–56.

Hayes, J., & Hopkins, A. (2014). *Nightmare pipeline failures: Fantasy planning, black swans and integrity management*. Sydney: CCH.

Hayes, J., & Maslen, S. (2015). Knowing stories that matter: learning for effective safety decision-making. *Journal of Risk Research, 18*(6), 714–726.

Klein, G. (1998). *Sources of Power: How People Make Decisions*. Cambridge, Massachusetts: MIT Press.

Main, B. W., & Frantz, J. P. (1994). How design engineers address safety: What the safety community should know. *Professional Safety, 39*(2), 33.

Middlehurst, R., & Kennie, T. (1997). Leading Professionals: Towards new concepts of professionalism. In J. Broadbent, M. Dietrich, & J. Roberts (Eds.), *The End of the Professions? The restructuring of professional work*. London: Routledge.

National Commission on the BP Deepwater Horizon Oil Spill and Offshore Drilling. (2011). *Deep Water: The Gulf Oil Disaster and the Future of Offshore Drilling*.

NTSB. (2012). *Pipeline Accident Report Enbridge Incorporated, Hazardous Liquid Pipeline Rupture and Release, Marshall, Michigan, July 25, 2010*.

Pidgeon, N., & O'Leary, M. (2000). Man-made disasters: why technology and organizations (sometimes) fail. *Safety Science, 34*, 15–30.

Polkinghorne, D. E. (1988). *Narrative Knowing and the Human Sciences*. Albany: State University of New York Press.

Schank, R. C. (1990). *Tell Me a Story: A New Look at Real and Artificial Memory*. New York: Maxwell Macmillan International.

Sullivan, W. M. (2005). *Work and Integrity: The Crisis and Promise of Professionalism in America*. San Francisco: Jossey-Bass.

Weick, K. E. (1993). The Collapse of Sensemaking in Organizations: The Mann Gulch Disaster. *Administrative Science Quarterly, 38*(4), 628–652.

Weick, K. E., & Sutcliffe, K. M. (2001). *Managing the Unexpected: Assuring High Performance in an Age of Complexity*. San Francisco: Jossey-Bass.

Wenger, E. (1998). *Communities of Practice: Learning, Meaning, and Identity*. Cambridge: Cambridge University Press.

Zabusky, S., & Barley, S. (1996). Redefining success: ethnographic observations in the careers of technicians. In P. Osterman (Ed.), *Broken Ladders: Managerial Careers in the New Economy*. New York: Oxford University Press.

Chapter 9
Doing What Is Right or Doing What Is Safe

An Examination of the Relationship Between Professionalization and Safety

Linda J. Bellamy

Abstract The relationship between professionalization and safety is examined from two angles. One is how professionals manage high risk by *doing what is right*, what the trusted professional does when applying their skills and expertise in controlling a hazardous technological system. The other is *doing what is safe*, control of safety by a formal system of regulation, rules, procedures, codes and standards which constrains behaviour to remain within specified boundaries of operation but which must still somehow allow sufficient flexibility of behaviour to adapt to the specifics of the situation. By looking at some of the aspects of accidents and lessons learned, issues in safety and professionalism are highlighted which could be important topics for developing safety competence, in particular for the management of uncertainty and unforeseen risks. The handling of uncertainties and the mitigation of cognitive bias, the development of tacit knowledge and the use of lessons learned from successful recoveries are potential learning opportunities for both types of professional for uncertainty management. The chapter first sets out to clarify the meanings in the title and how they are different and possibly even conflicting in the approach to safety. This theme continues in examining how these issues are reflected in accidents and lessons learned. Finally, some possible ways forward are identified.

Keywords Tacit knowledge · Cognitive bias · Uncertainty management

L.J. Bellamy (✉)
White Queen Safety Strategies, Hoofddorp, The Netherlands
e-mail: linda.bellamy@whitequeen.nl

© The Author(s) 2018
C. Bieder et al. (eds.), *Beyond Safety Training*, Safety Management,
https://doi.org/10.1007/978-3-319-65527-7_9

85

9.1 Introduction

Two types of professional activity relate to the control of the hazards, namely *doing what is right* and *doing what is safe*. These labels simplify the more complex understanding of what is meant in these two areas. *Doing what is safe* refers to the human-involved activity of safety control which is largely governed by rules and procedures, with criteria for how things are done and what the outputs of the processes should be. Keeping within the boundaries of the formal system is achieved by compliance. However, when formal control systems break down or are ambiguous or wrong and in any case are never complete in covering variations in the working environment, those at the sharper end are left to resolve the control dilemmas in other ways.[1] This is labelled here as *doing what is right* because it depends on knowing what is the right thing to do. It is similar to what Colas (1997) described as the positive contribution of human reliability:

> (...) participation in building safety, understanding of situations, the unavoidable "operational adaptation" in areas the rules do not cover, good habits and the real situations to which they apply, depending on the specificities of those situations.

9.2 Doing What Is Right

For *doing what is right* there is a dependence upon professionalism, a particular kind of expertise for which a person can be trusted to perform both according to established high standards as well as ethically. Professional bodies, like the Institution of Chemical Engineers for example, emphasise the importance of public trust in the professional. There is a normative value system of professionalism in work which is expected at the micro level in individual practices in the workplace. Membership of professional bodies depends upon adherence to accepted knowledge and competence standards and codes of practice.

Holtman (2011), a medical professional, says that expertise—specialised skills and knowledge—is the foundational element of professionalism. Specialised knowledge is often used to distinguish a profession from an occupation; an occupation is considered to rely more on craft skills even though craft skills are also required by professionals. According to Vitale (2012, 2013) professionalization of work has intellectualised occupations and transformed them into knowledge-based professions. This separation from craftsmanship is seen as an inhibitor of inter-professionalism with the creation of in- and out-groups. In the field of safety Almklov et al. (2014) have suggested that the knowledge element produced by

---

[1]For an extreme example of this see Fink (2013) describing the life and death dilemmas faced by staff at the Memorial hospital in New Orleans after Hurricane Katrina hit, taking out essential support systems and eliminating options for recovery.

scientists, which journals like Safety Science purport to communicate, may actually contribute to undermining local and system specific personal expertise and hence the unwritten craft skills associated with craftsmanship. Top professionals managing high risks recognise the importance of these skills:

> ...thanks to your experience you will have the right reflexes and will act in a way in which you will find the right answer for situations that are extremely difficult. *Professional Alpinist & Adventurer* (Van Galen and Bellamy 2015).

This kind of knowledge is also called "tacit" knowledge, a term introduced by Polanyi (1958) to distinguish it from propositional knowledge in that it cannot be communicated by language or mathematics. The person knows how to do something but cannot explain how it is done. Maslen (2014) has looked at building safety knowledge among new engineers in the pipeline industry. She provides evidence that this kind of tacit knowledge associated with craft skills is important and not easily passed on. It has to be developed in the workplace from experience and mentoring.

The skills and expertise expected of the professional are called upon when there are uncertainties in how to respond. There are unforeseen risks and surprises in practice which are not immediately covered by a normative risk management system. This demands a certain amount of flexibility which may be difficult to define in a system focused only on *doing what is safe*.

9.3 Doing What Is Safe

The safety of any activity must be assured for it to be acceptable. The safety of an activity is like a piece from a cake with many layers built up over time as contexts and knowledge change and develop as shown in Fig. 9.1. This model was originally designed to provide a scenario based approach to inspection and auditing of chemical plants (Oh and Bellamy 2000; Bouchet 2001). Any activity cuts through many perspectives: the knowledge of hazards and how to control them, the management of the processes which ensure integrity of the system, the human role at the different levels of control and, not shown, the environment of the system itself. The integrity of the technical system is at the core of the model. If this functioned without any further need for intervention there would be no requirement to manage it but this is rarely if ever the case. Maintaining the integrity of the technical system, keeping within the boundaries of the formal systems—codes, standards, rules, regulations, procedures, best practices, safety margins etc., are defined here as *doing what is safe*. It is what the public might expect, to be protected in some definable way. Recently, new efforts have been made to grapple with the complex nature of the problem. For example, Le Coze (2013) provides a "sensitising model" covering multiple perspectives (technological, organisational, psychological, cultural, sociological, political) emphasising their intertwined nature at a micro-meso- and macro-level.

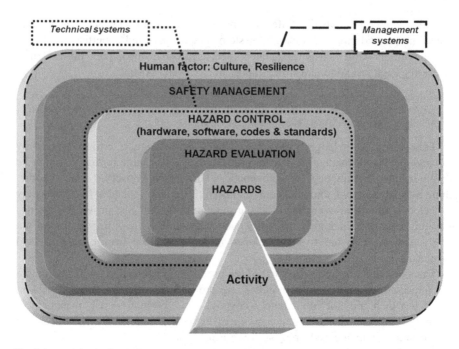

Fig. 9.1 A "piece of cake": an activity associated with a hazardous technology contains the ingredients of the socio-technical system as understood and regulated

The purpose of choosing the words *doing what is safe* is to also link to the role of safety professional. The professionalization of safety was originally associated with enforcing legislation directed at accident prevention and was not integrated with the work. Factory inspectors for this purpose originated in the mid-19th century (Hale and Harvey 2012; Mohun 2013) and were the first safety professionals. Later, safety officers were appointed to ensure compliance in companies. Today the safety professional's role has expanded beyond compliance in the evaluation and control of the hazards. Regulation has developed with an increasing focus on risk management like the layers in Fig. 9.1. The safety professional's role is much concerned with hazard identification, assessing the risks and accident investigation when hazard control fails. Problem solving, advising, training and communicating also need to be considered along with appropriate technical knowledge. Wybo and Wassenhove (2016) distilled four main activities of HSE professionals as: regulatory compliance, design and operation of the HSE management system, risk and accident analysis, and emergency and crisis management. Although regulatory compliance is now regarded in the literature as low importance, compliance with laws, codes and standards is considered fundamental to risk management. In the Netherlands, investigated serious accident reports and major hazard loss of containment reports are analysed with the tool Storybuilder (Bellamy et al 2013, 2014). The extensively analysed data indicate however that only around half the accidents are found to have legal breaches, whether occupational or loss of containment accidents (previously unpublished data).[2] Compliance does not mean accidents are always prevented.

9.4 Problems

Doing what is safe may contribute to causing accidents. The Three Mile Island (TMI) nuclear accident in 1979 (Kemeny 1979) is an example where following the procedures actually made things worse. TMI was a turning point for highlighting the importance of operator support (interface design, procedures, training) in handling complex technologies. After the Chernobyl accident in 1986 and the conclusion that operators had been violating *doing what is safe,* the concept of Safety Culture was introduced (IAEA 1991). It was emphasised that safety as a collective attitude should be given number one priority, something that goes beyond procedure following and which requires a questioning attitude, stopping and thinking, and communicating with others. Later however design deficiencies were revealed that suggested it was the Chernobyl plant design that was at fault as well as arrangements for presenting important safety information to operators (IAEA 1992):

> Certain actions by operators that were identified in INSAG-1 as violations of rules were in fact not violations (p.24).

The system of *doing what is safe* can always be used to blame operators rather than support them.

The reality is that professionals engaged in operating a system have multiple performance goals to achieve:

> Resources will always be limited. So you need to think carefully, "when is enough, enough?" ... You can optimize, and plan and prepare to perfection, but that means you never move. So you need to have some courage and say, okay let's go, now we're going to execute and do. *Process manager, Petrochemical Industry* (Van Galen and Bellamy 2015).

Ale (2005) says that courageous acts should be preceded by consulting "a morbid pessimist like a risk analyst". While this could be formalised and turned into a matter of routine, even so there is still a need to push on. Besides that, not all risks can be foreseen. When there are many uncertainties, and so an incomplete model of knowledge and risk, and there are high pressures for the day to day operations to continue, balancing safety and production becomes what one manager called "top level sport": being able to reduce the human error statistics in decision making without the performance going down too much (Van Galen and Bellamy 2015).

---

[2]Currently there are around 25,000 occupational accidents (around 44% with no breach of the law found) and 270 loss of containment accidents (around 50% with no breach of the law found) analysed in the Dutch tool Storybuilder. The software and databases can be freely downloaded at: http://www.rivm.nl/en/Topics/S/Storybuilder.

Another perspective on this is "mindfulness" (Weick and Sutcliffe 2007). Mindful organising is a collective behavioural capability to detect and correct errors and adapt to unexpected events. In *doing what is right* one would expect professionals to be mindful in attention to detail and quality such as checking their own work, to be observant to the possibility of making and correcting errors as part of successful recovery and to be aware of possible biases in thinking.

The unexpected is about not knowing. For example, Kletz (2009) in his book "What Went Wrong" has a chapter entitled *I Did Not Know*. The Flixborough accident in the UK in 1974 (Department of Employment 1975) was an explosion at a chemical plant that killed 28 people. It is an example of a major hazard accident involving not knowing what you do not know; in order to by-pass a failed reactor to get production going again, no-one thought it was anything other than a routine plumbing job. This resulted in the court of enquiry recommendation that the training of engineers should be more broadly based and that all engineers should learn at least the elements of other branches of engineering (*Recommendation 210 (ii)* (Op.cit.). This was an exceptional situation. However, routine non-compliances may also occur where professionals work. Take pilots, for example. Apparently 97% of unstabilised approaches to a landing continue to be flown contrary to airline Standard Operating Procedures (Go-around safety forum 2013). The pilot may think that making a go-around instead of continuing the landing is more risky or that the go-around criterion is overly stringent. It is argued from the failure data that non-stabilised approaches increase the number of accidents like runway excursions and that it is the inadequate situational awareness of pilots that is causing an inaccurate risk assessment. Here is a conflict; where to draw the line between compliance and professionalism?

> Standard work methods are important for standard work, but when things are happening that are not standard and the organisation only is built around these standard procedures, you will lack people that have the possibility to think thoroughly *Production assistant, Steel-making industry* (Van Galen and Bellamy 2015).

An example of this is given by the failure of workers to "drop your tools" (Weick1996). Weick describes two separate incidents where firemen escaping wildland fires failed to drop their heavy tools despite the fact that they needed to do so in order to quickly escape the threat. Failing to outrun the flames, many perished. This lack of flexibility is brought about by adaptation and learning. Conversely, there can be too much potential flexibility of behaviour when it has not been adequately constrained by procedures and learning. In a train derailment accident, lack of skills in handling tools and equipment was an underlying cause in a failure to identify defective railway points (RAIB 2009). In-house competence training did not include hands-on use of the tools or equipment on the infrastructure for which it was intended for use. Worn rails would have been needed to demonstrate a failed switch; the trainer used a computer presentation.

Although knowledge, expertise and experience from doing may be beneficial in finding the right solution to risk problems under uncertainty, there are important flip sides to experience, such as overconfidence (Roberto 2002). The human perception of and response to uncertainty is the subject of psychological debate (Smithson 2008) with a literature beyond coverage here. Cognitive models explaining judgement and decision-making outcomes under uncertainty have received attention, such as Endsley (2000) on situation awareness, a dynamic relationship with the cues from the environment for knowing what is going on around you. Another area concerns the limited availability of cognitive resources. For example, Hollnagel (2009) considers that when these cognitive resources are challenged, such as under time pressure, it results in an efficiency-thoroughness trade off (ETTO) in performance. Another camp (Tversky and Kahneman 1973, 1974; Kahneman 2011) suggests that even when it is within a person's capacity to reason correctly they still use heuristics (mental shortcuts) and cognitive biases (tendencies to come to a wrong conclusion). When competent professional people make discretionary decisions they are not necessarily free of bias.

Another bias occurs when safety performance is measured in hindsight. Hindsight bias (the wisdom of hindsight) is a projection of new knowledge onto past events as though that knowledge were available at the time. Fischoff (1975) suggests that hindsight bias undermines our ability to appreciate the nature of uncertainty and surprises and to learn from them to improve the system. Lessons learned derived from an analysis of accidents are a form of hindsight bias, usually towards a normative model of safety. They are intended to encourage people to think of safety in their own work context and usually specified in terms of failures e.g. see ARIA[3] accident database of technological accidents (BARPI[4] n.d.) and lessons learned from the eMARS[5] major accidents database (European Commission n.d.). Little lessons learned attention is given to successful recoveries (Resilience Success Consortium 2015). One aspect of biased thinking is that it can lead to depreciation of successful recoveries by attributing them to luck. In fact, professionals with a role in dealing with high risk tend not to show much interest in learning from success, and even suggest that this could be potentially dangerous because of certain heuristic traps (Van Galen and Bellamy 2015). Little if any lessons learned attention was given to, for example, the fact that unarmed but aware and trained persons were able to thwart an armed terrorist on the Thalys train travelling from Amsterdam to Paris on 21 August 2015. After a brief interest in the heroism of these persons who applied their professional knowledge and skills in an unexpected context, attention focused instead on the weaknesses of the system. Positive recoveries are rarely analysed for factoring into risk assessment models and are easily dismissed as luck or heroism, with a consequent absence of data that would enable their effects to be quantified. Take an example—the ditching of a passenger flight in the Hudson River in the US after bird strike caused a complete lack of thrust. Successful recovery, it was concluded, mostly resulted from "a series of fortuitous circumstances" (NTSB 2010, p.79).

---

[3]Analysis, research and information on accidents (editors' note).

[4]Bureau for Analysis of Industrial Risks and Pollutions (editors' note).

[5]Major Accident Reporting System (editors' note).

9.5 Conclusion and Solutions

Doing what is safe and normative safety management is about constraining human performance. Human beings can follow such a system but this only deals with foreseen risks. In this system the human is regarded as a component that could deviate from the norms and the system is then considered vulnerable to errors and violations when there is human interaction.

The human potential for safety rests with the professional in *doing what is right*. Here it is an uncertainty management system that is required that allows human beings the flexibility to adapt and deal with the unforeseen risks as they arise, the unexpected events and surprises that can occur. It is through developing the professional as part of an uncertainty management system that the human potential for safe performance can be enhanced. It is only recently that this is being recognised. In Fig. 9.1 a new layer was added, the human factor, and explicitly resilience and safety culture. It is this area that is the important part of the uncertainty management system.

A number of problems have been identified:

- Treating safety as if it is only about keeping to norms and not including other aspects of safe behaviour associated with the management of uncertainty misses the opportunities to develop and use professionals for thinking about and dealing with the unforeseen.
- Knowledge and experience are important in uncertainty management yet developing experience takes time; there is the need to acquire craft skills/tacit knowledge which enhances the capacity to deal with the unexpected but this is not the type of knowledge that can be written down—it has to involve doing.
- When faced with uncertainty, it is a challenge to optimise the human cognitive functions when also having to deal with resource constraints (like limited time, tools, information), conflicting goals (like safety constraints versus the need to push on) and cognitive biases (such as overconfidence).

Some solutions to these problems might be found in changing the way we think about safety. Already there is more emphasis on integration between *doing what is safe* and *doing what is right*. Jørgensen (2015) says that accident research shows that safety must be integrated with the whole enterprise and function on all levels of management. It is necessary to make safety a part of professionalism without making it a separate issue; it should be an integrated part of the right way to do things and aligned to the mission of the organisation. This means that safety is not picked out as an independent separate item. Good operators will be attentive to everything and have the judgement and decision-making skills for dealing with uncertainties.

However, the necessary experience for certain jobs may take years to develop. Researchers are recognising the value of considering ways to communicate the kind of knowledge that develops from experience. For example, Podgórski (2015) examines the use of tacit knowledge in occupational safety and health management systems in the context of knowledge management. He finds that the methods of information transfer about knowledge rooted in the working context are not well studied, although there are some promising solutions like hazard awareness apprenticeships in the field and the use of virtual realities. Maslen (2014) identifies that for new engineers learning from experience can be facilitated by actual participation early on the job by being given responsibilities like being in charge of a project instead of just observing how things are done.

Professionals should also have good tools. Ideally the tools of hazard control, unlike the earlier mentioned heavy tools of the firemen (Weick 1996), should help to buy time in situations of threat and uncertainty and not limit it. Time provides the opportunity for the gathering of multiple perspectives and expertise for reducing the uncertainty of a successful outcome and deciding actions when things deviate. In making decisions in the face of uncertainty, professionals managing high risk have pointed out the advantages of thinking and deciding together, of sharing observations, knowledge and experience and of having a group of opinions to find the right path through the uncertainties (Van Galen and Bellamy 2015). Even that may be subject to biases such as groupthink (Janis 1972); biases are always there compromising successful outcomes. Learning about biases is a process lacking in training resources although there are recommendations for debiasing strategies in the literature such as simulation, thinking about thinking, or forcing consideration of alternatives (Croskerry 2003). Professionals should be aware of biases and how to mitigate their adverse effects as well as enhance their good ones as efficient strategies such as when time is limited. The role of safety professionals is very much in the normative camp. They are the watchmen and enforcers when the risks are foreseen. But safety professionals cannot deal with the uncertainties when they lack the specialist professional skills and experience of the subject expert. They cannot be surrogates for mindfulness in this respect. An important role however could be as a devil's advocate or "morbid pessimist" or at least to facilitate the acquisition of such a resource.

The process of generating increasing mindfulness is, as Weick and Sutcliffe (2007) point out, not more equipment or more training or more of the old strategies. The thinking in this chapter leads to the conclusion it is more about how to:

- Recognise and develop uncertainty management systems for handling unforeseen risks.
- Integrate safety with the other goals of the system in professional training, identifying how multiple goals can be achieved through *doing what is right* and not just by *doing what is safe*.
- Learn to recognise and reduce uncertainty and bias, with the need to develop resources to support this and to make use of connections to people with other perspectives. This has an organising aspect of uncertainty management for enabling this connectivity and communication.
- Improve and speed up acquisition of tacit knowledge through using new possibilities for apprenticeship, such as to virtual masters, and organisational solutions like giving new recruits early opportunities to be responsible for work.

- Improve safety as an increased awareness for successes in past recoveries where there were uncertainties associated with changes, deviations or near misses, using lessons learned about successes as part of investigation and learning about how to better anticipate the unknown.

References

Ale. B.J.M. (2005). Living with risk: a management connection. *Reliability Engineering and System Safety, 90,* 196–205.

Almklov, P.G., Rosness, R., & Størkersen, K. (2014). When safety science meets the practitioners: Does safety science contribute to marginalization of practical knowledge? *Safety Science, 67,* 25–36.

BARPI (n.d.). *The ARIA database.* Bureau d'Analyse des Risques et Pollutions Industriels. Retrieved from: http://www.aria.developpement-durable.gouv.fr/about-us/the-aria-database/?lang=en.

Bellamy, L.J., Mud, M., Manuel, H.J. & Oh, J.I.H. (2013). Analysis of underlying causes of investigated loss of containment incidents in Dutch Seveso plants using the Storybuilder method. *Journal of Loss Prevention in the Process Industries,* 26, 1039–1059.

Bellamy, L.J., Manuel, H.J. & Oh, J.I.H. (2014). Investigated Serious Occupational Accidents in The Netherlands, 1998–2009. *International Journal of Occupational Safety and Ergonomics (JOSE), 20* (1), 3–16.

Bouchet, S. (2001). *Analyse des risques et prévention des accidents majeurs (DRA-07). Présentation des méthodes d'inspection TRAM, NIVRIM et AVRIM2.* INERIS, Rapport intermédiaire d'opération d, Prise en compte de l'influence des barrières de sécurité dans l'évaluation des risques. Retrieved from: http://www.ineris.fr/centredoc/28.pdf.

Colas, A. (1997). *Human factor in the nuclear industry.* Paper presented at the specialist meeting on operational events, Organisation for Economic Co-operation and Development/Nuclear Energy Agency, Chattanooga, TN. Retrieved from: http://www.iaea.org/inis/collection/NCLCollectionStore/_Public/44/026/44026252.pdf.

Crosskerry, P. (2003). The importance of cognitive errors in diagnosis and strategies to minimise them. *Academic Medicine,* 78 (8), Aug 2003.

Department of Employment (1975). *The Flixborough disaster. Report of the Court of Inquiry. Formal Investigation into Accident on 1st June 1974 at the Nypro Factory at Flixborough.* London: HMSO.

Endsley, M.R. (2000). Theoretical underpinnings of situation awareness: a critical review. In Mica R. Endsley & Daniel J. Garland (Eds.), *Situation awareness analysis and measurement* (pp. 3–32). Mahwah, NJ: Lawrence Erlbaum Associates.

European Commission (n.d.). *Learning lessons from accidents to prevent future accidents.* Retrieved from: https://minerva.jrc.ec.europa.eu/en/content/minerva/f4cffe8e-6c6c-4c96-b483-217fe3cbf289/lessons_learned_from_major_accidents.

Fink, S. (2013). *Five days at Memorial: Life and death in a storm-ravaged hospital.* London: Atlantic Books.

Fischoff, B. (1975). Hindsight foresight: the effect of outcome knowledge on judgment under uncertainty. *Human Perception and Performance,* 1, 288–299. Reprinted in *Quality and Safety in Health Care 2003, 12* (4), 304–311. Retrieved from: http://www.ncbi.nlm.nih.gov/pmc/articles/PMC1743746/pdf/v012p00304.pdf.

Go-Around Safety Forum (2013*). Findings and conclusions,* 18 June 2013, Brussels. Retrieved from: http://www.skybrary.aero/index.php/Portal:Go-Around_Safety.

Hale, A. & Harvey, H. (2012). *Certification of safety professionals: emerging trends of internationalisation.* Proceedings of the 6th International Conference of Working on Safety Network "Towards Safety Through Advanced Solutions", Sopot, Poland, 11–14 September 2012.

Hollnagel, E. (2009). *The ETTO Principle: Efficiency-Thoroughness Trade-Off.* Ashgate Publishing.

Holtman, M.C. (2011). Paradoxes of professionalism and error in complex systems. *Journal of Biomedical Informatics*, 44, 395–401.

IAEA (1991). *Safety culture.* Safety series No. 75-INSAG-4. Vienna, Austria: IAEA.

IAEA (1992). *The Chernobyl accident: updating of INSAG-1.* Safety series No. 75-INSAG-7. Vienna, Austria: IAEA.

Janis, I. L. (1972). *Victims of groupthink.* Boston: Houghton Mifflin.

Jørgensen, K. (2015). Integration of safety in management in tasks in onshore transport SME's. pp. 50– 62 in *Proceedings of the 8th International Conference Working on Safety (WOS)* 23-25 Sept, Porto, Portugal.

Kahneman, D. (2011). Thinking fast and slow. New York: Farrar Straus & Giroux. (Paper back 2012 Penguin Books).

Kemeny, J.G. (1979). Report of the President's Commission on the accident at Three Mile Island. John G. Kemeny, Chairman. Retrieved from: http://www.threemileisland.org/downloads/188.pdf.

Kletz. T. (2009). *What went wrong* (5th edition). Amsterdam: Elsevier.

Le Coze, J-C. (2013). Outlines of a sensitising model for industrial safety assessment. *Safety Science, 51*(1), 187–201.

Maslen, S. (2014). Learning to prevent disaster: An investigation into methods for building safety knowledge among new engineers to the Australian gas pipeline industry. *Safety Science, 64*, 82–89.

Mohun, A. (2013). *Risk: Negotiating Safety in American Society.* Baltimore: John Hopkins University Press.

NTSB (2010). *Loss of Thrust in Both Engines After Encountering a Flock of Birds and Subsequent Ditching on the Hudson River US Airways Flight 1549 Airbus A320-214, N106US Weehawken, New Jersey January 15, 2009.* Aircraft Accident Report. NTSB/AAR-10/03 PB2010-910403, National Transportation Safety Board, USA.

Oh, J.I.H. & Bellamy L.J. (2000). AVRIM2: *A holistic assessment tool for use within the context of the EU Seveso II directive.* Seveso 2000 conference, 22–23 June, Bordeaux, France. http://whitequeen.nl/papers_files/Avrim2bordeaux.pdf.

Podgórski, D. (2015). The Use of Tacit Knowledge in Occupational Safety and Health Management Systems. *International Journal of Occupational Safety and Ergonomics, 16*(3), 283–310.

Polanyi, M. (1958). *Personal Knowledge: Towards a Postcritical Philosophy.* London: Routledge and Kegan Paul.

RAIB (2009). *Derailment near Exhibition Centre station, Glasgow, 3 September 2007.* Rail Accident Investigation Branch, Department for Transport, Report 04/2009, February 2009. https://www.gov.uk/raib-reports/derailment-near-exhibition-centre-station-glasgow.

Resilience Success Consortium (2015). *Success in the face of uncertainty. Human resilience and the accident risk bow-tie.* Final report of a European SAF€RA project, March 2015. www.resiliencesuccessconsortium.com/resources Retrieved from: http://dx.doi.org/10.13140/RG.2.1.3670.1046.

Roberto, M.A. (2002). Lessons from Everest: the interaction of cognitive bias, psychological safety, and system complexity. *California Management Review, 45*, 136–158.

Smithson, M. (2008). Psychology's ambivalent view of uncertainty. In Gabriele Bammer and Michael Smithson (Eds.), *Uncertainty and Risk: Multidisciplinary Perspectives* (pp. 205–217). London: Earthscan Publications Ltd.

Tversky, A., & Kahneman, D. (1973). Availability: A heuristic for judging frequency and probability. *Cognitive Psychology, 5*, 207–232.

Tversky, A., & Kahneman, D. (1974). Judgment under Uncertainty: Heuristics and Biases. *Science, 185* (4157), 1124–1131.

Van Galen, A. & Bellamy, L.J. (2015). *Resilience Case Studies*. Annex B to Resilience Success Consortium main report: Success in the face of uncertainty. www.resiliencesuccessconsortium. com/resources, February 2015. Retrieved from: http://dx.doi.org/10.13140/RG.2.1.4063.3209.

Vitale, JE. (2012). An Essay on the Division between Craft-Based and Knowledge-Based Professions as an Inhibitor of Interprofessional Healthcare Education and Practice, Part I. *Health and Interprofessional Practice, 1*(3), eP1026. Retrieved from: http://dx.doi.org/10. 7772/2159-1253.1026.

Vitale, JE. (2013). An Essay on the Division between Craft-Based and Knowledge-Based Professions as an Inhibitor of Interprofessional Healthcare Education and Practice, Part 2. *Health and Interprofessional Practice, 1*(4), eP1047. Retrieved from: http://dx.doi.org/10. 7772/2159-1253.1047.

Weick, K.E. (1996). Drop your tools: An allegory for organizational studies. *Administrative Science Quarterly, 41*(2), 301–313.

Weick, K.E. & Sutcliffe, K.M. (2007). *Managing the unexpected: Resilient performance in an age of uncertainty* (2^d edition). San Francisco, CA: Jossey-Bass.

Wybo, J-L. & Wassenhove, W.V. (2016). Preparing graduate students to be HSE professionals. *Safety Science, 81*, 25–34.

Chapter 10
Industrial Perspective on the Seminar: The Viewpoint of a Mining Expert

Jonathan Molyneux

Abstract Based on his extensive expertise in the mining industry, Jonathan Molyneux raises the issue of the importance of operational experience, besides acquiring formal safety qualifications, to improve safety performance in high-hazard industries. He highlights the paradox by which the influencing aspect of the work of "safety professionals" as valued advisors is somehow challenged by the fact that they have to meet the compliance agenda and are therefore sometimes perceived by shop floor staff more as a "procedure-police" than as coaches. Integration versus differentiation with safety improvement strategies tailored for specific local contexts is also discussed.

Keywords Trade skills · Compliance to standards · Balance between goals and incentives

Kudos to FONCSI for bringing together a powerhouse of experience to wrestle with a provocative aspect of industrial safety improvement. Delegates joined from a cross section of disciplines (industrial psychologists, industry safety practitioners and business improvement advisors) and industry sectors (chemicals, aviation, oil and oil services, medical and mining).

The focus of the workshop[1] was the "Professionalization of Safety" and the extent to which this might hold the key to advance injury and fatality reduction—a challenge all workshop attendees agree remains necessary and urgent, especially in high-hazard sectors, and despite improvements over recent years. A number of the participants had been involved in the Toulouse catastrophe in 2001[2] and had first-hand experience of the investigations and consequences. All participants had in

---

[1]The two-day international workshop mentioned in the preface, organized by FonCSI in November 2015 and highlight of the project that led to this book (editors' note).
[2]The AZF chemical factory exploded in Toulouse, France, on 21th September 2001 (editors' note).

J. Molyneux (✉)
ERM, London, UK
e-mail: jonathan.molyneux@erm.com

© The Author(s) 2018
C. Bieder et al. (eds.), *Beyond Safety Training*, Safety Management,
https://doi.org/10.1007/978-3-319-65527-7_10

some way devoted a substantial section of their career to the prevention of similar incidents, as well as less headline-grabbing, but all too frequent events that result in the loss of life. Examples of experience which was shared by participants included:

- The design and roll-out of a safety improvement program across the worldwide operations of an oil services company. This highlighted the value of gaining investment support from a company's executive team, strong branding and communications, a dedicated project team and a well-defined and simple to understand set of tools and actions which, when adopted by the workforce, can create changes in behaviour and safety performance. This programme had produced marked performance improvement; and,
- The development and execution of a performance improvement intervention which had been run at 50 individual mine sites around the world. This work had revealed that the improvement challenges at each site, while bearing many similarities, were different. The focus had been on building alignment amongst the mine management team and persuading them to work together in a co-ordinated way on a small number of underlying aspects of how their workforce think and manage work activities. This approach had also demonstrated strong results.

It became apparent that each of these case examples had benefited from thought and investment, each had delivered safety performance improvements—and yet, all presenters acknowledged that much work remained to be done to reach a point of "Zero Harm" (meaning zero injuries). This topic itself provoked some discussion; the term has been adopted widely in a number of industries as a means of conveying the intent of safety programmes. While some participants saluted the good values the term embodies (that *no* injury is acceptable, and that *all* injuries are preventable) and the power this has in challenging the mind-sets of managers and workers, others pointed to the idealism of the ambition. This author has found that improvement efforts needed to be more obtainable and focused on specific issues and root causes influencing fatalities and incidents which have the potential to result in life changing injuries.

There was also some debate that challenged the widely endorsed safety incident ratio model (Bird's Pyramid). In mining this belief had historically led to safety improvement efforts focusing on the prevention of high-frequency/minor injuries with an assumption that work at this level would contribute to changes in behaviour which would ultimately reduce the probability of a fatality. More recent thinking is that this course of intervention is not as effective as applying more deliberate focus to the specific pre-cursors of fatal events themselves, which we now recognise are often different to the higher-frequency areas that have attracted attention when applying the pyramid model.

The concept of professionalising safety provoked two interpretations, each one contributing positively to the debate.

The first interpretation was the proposition that raising the level of professional standing, qualifications and perhaps staffing levels of people with roles devoted to safety improvement would equip companies with a stronger, more influential and effective force for improvement. Participants pointed out that there are already well established qualifications and professional development tracks for safety—they also noted the recent trend of appointing people with operational experience (rather than formal safety qualifications) to senior safety function roles. These two perspectives seemed to highlight the potentially divergent challenges of having the:

- expertise to diagnose and define improvement strategies; and,
- ability to influence change and win the support of line managers who typically hold ultimate sway over the realities of work on the shop floor.

Could this discussion highlight the need for a team approach, and perhaps some caution over developing too narrow a professional profile?

In the author's experience from mining the role of safety professionals at operational level is typically divided between ensuring compliance to regulatory and corporate standards on the one hand, and influencing behavioural performance improvement on the shop floor (via leadership behaviours at managerial levels). A common challenge is that "professional" safety people are drawn to the technical demands of the compliance agenda, especially when audit results are typically seen as an indirect indicator of their personal performance. The paradox is that the influencing aspect of their work, especially when focused on a well-shaped improvement strategy, is usually more central to incident reduction, but it is an altogether more challenging, sometimes even abrasive activity. In mining we often find operations see their safety function colleagues as procedure-police, rather than as valued advisors and coaches; might improved "professionalization" be best focused on the non-technical aspects of the functional team's skill sets?

A second interpretation of the professionalization concept looked at how to integrate safety into operations. That is to say, the proposition that rather than concentrating investment on safety function professionals, a route to major improvement may lie in the integration of safety understanding, safety thinking and safety management into the mainstream of business and operations. This approach would encourage business decision makers to achieve a more considered balance between the likelihood of achieving their target commercial outcomes AND achieving safety target outcomes. Some workshop delegates felt that in many industries, this is not currently the case—another view was that many decision makers DO appreciate the balance, but are compelled by commercial pressures to make decisions which contribute to workforce risk, and they rely on the operational dexterity of their people to absorb, resolve and deliver (often referred to as "re-silience"). This approach would suggest that the operational managers and front line supervision should be the targets for professionalization. With such an enhanced safety orientation, they might more naturally balance production and maintenance decisions with risk management, or better still, to integrate risk management into routine operational thinking and procedures. This way they would be better equipped to provide the right guidance to their workers and ensure that workers are genuinely set up for success in the workplace.

Towards the end of the workshop a provocative question was raised by one of the faculty; many industry executives have approved multi-year investments in safety training—yet they feel these investments have not been productive. They ask how they can yield a stronger safety performance improvement for their investments.

This question prompted an animated debate. The delegates remarked on the range of training options which might contribute to safety performance; new employee induction training, safety leadership training for operational managers, the use of risk control techniques at different levels in the organisation—and non-safety specific training in core trade skills which support the execution of good work with good tools leading to good results.

This author's energies have always been focused on safety performance *improvement* in mining. In most situations, diagnostics and solution development have highlighted training as a relatively *low impact* mechanism for creating performance change. It is typically an element of the corporate apparatus which supports the status quo rather than driving material improvement—aspects of training are often necessary for compliance and in mining we have seen situations where related training programs are them too cumbersome for the operations to keep up with, and participant feedback has revealed the sessions themselves to have made little lasting impact on how people perform their work and the decisions they make.

In mining, experience has demonstrated that to achieve sustainable performance improvement senior management and the architects of safety programs should consider:

- The intertwining of what makes humans the creative creatures we are with the balance of goals and incentives we put before workers; the pre-cursors to incidents typically lie with how the organisation is routinely run and the level of acceptance of hazardous activities that this incubates, rather than with the shortcomings or lack of judgement of the unfortunate individuals who so often appear to have "ignored the rule";
- That each operation is unique; unique risk profile, operating culture and leadership dynamics. This means that creating sustainable improvement requires a tailored approach that targets the most appropriate improvement levers applicable to the individual operation—a local safety improvement strategy;
- Improvement strategies need to change how people think about their work, how teams work together and the decisions that individuals make; so they need to be led through a deliberate coalition between line managers and supervisors on the shop floor and their safety function advisors, rather than via a sanitised training setting; and,
- Interventions need to cut to the heart of how work is designed and scheduled, an increased sophistication in how teams identify and control the hazards in their work, and centrally, what level of exposure operational leaders are prepared to accept for their people.

Chapter 11
How to Deal with the Contradictions of Safety Professional Development?

An Organizational Approach Based on Discussion

Benoit Journé

Abstract Companies around the word currently ask their employees to behave and work as "professionals". To be a "pro" has become a managerial leitmotiv that promotes an ideal image of employees based on the highest levels of performance, rationality, responsibility and reliability, especially in the domain of risk industries and safety management. This is typically the vision that managers promote when they decide that "failure is not an option". Hence, the development of employee professionalism appears to be a very legitimate and neutral objective that should be at the core of the functions of the Human Resource Management. In every big company, many resources of all kinds have been invested to design and implement increasingly sophisticated training programs for professional development and to engage managers and HR's departments. Unfortunately, these efforts have not produced the expected pay-offs in terms of safety performances and this disappointing performance raises several questions and problems. This chapter addresses them and suggests that some of the basic assumptions and images companies currently use to manage professionalism and professionalization are misleading because they over-simplify their nature. In other words, the notions of performance, rationality, responsibility and reliability that are associated with professionalism are in fact totally oriented towards compliance with formal procedures and rules. In some ways, the "professional" is seen as the perfect employee that never makes errors, never fails and never complains. In fact, this vision is purely behavioral (i.e. exclusively based on personal behaviors) and neglects the social and the political roots of professional skills and competencies. This chapter (1) identifies some of the main tensions and contradictions that are tightly linked to the notion of professionalism and (2) suggests how to actively manage these contradictions and explores new ways to develop professionalism in risk industries.

Keywords Contradictions · Discussion spaces · Safety practices

B. Journé (✉)
Université de Nantes, LEMNA, Nantes, France
e-mail: benoit.journe@univ-nantes.fr

C. Bieder et al. (eds.), *Beyond Safety Training*, Safety Management,
https://doi.org/10.1007/978-3-319-65527-7_11

11.1 The Managerial/Bureaucratic Approach Versus The Profession/Trade Approach

We suggest that the main problems come from the political opposition between a managerial/bureaucratic approach to professionalism on the one hand, and a profession/trade approach, on the other hand. The managerial/bureaucratic approach aims to develop the performances of the organization through a process of rationalization based on formal rules, guidelines and best practices. Better rules mean better efficiency as well as better safety for the organization. Therefore, the good practitioner, (i.e. the "pro"), is supposed to be a "perfect" employee that follows procedures and complies with the rules set out and implemented inside the organization. Unfortunately, this is not exactly the reality, even in risk industries, at least for two reasons.

The first reason is pragmatic. The managerial/bureaucratic approach promotes a behavioral vision of professionalism based on the "expected good behaviors" determined by the designers of the technical system as well as the managers of the organization. The problem is that safety comes also from the ability of the practitioners to cope with unexpected situations and events. In such cases, there is no "expected" behavior. This one may be defined *ex post* from the final outcome of the situation, but during the real-time activity, it is the responsibility of the professional to adjust and adapt its behavior as the situation develops, including in unexpected ways.

The second reason is more political. The progressive development of the managerial/bureaucratic approach since the end of the 19th Century was a political fight against the traditional vision of professionals as skilled practitioners strongly socialized in professional groups that were governed in compliance with their own norms and rules, coming from the outside the organizations they worked in. The classic power balance between the company owners and their employees shifted dramatically with the emerging "professionalization" of managers. This new kind of professional was highly trained in management techniques through MBA programs. They were hired by owners to develop the efficient model of large modern companies based on a bureaucratic rationalization of work and organization as defined by Frederick Taylor (1911) and Henri Fayol (1916). This constituted the "managerial turn" described by Berle and Means (1932) with the rise of "managerial firms" and the fall of the traditional "entrepreneurial firm". In managerial firms, owners stay outside the company. They are simply shareholders that have delegated all the organizational responsibilities to the professional top managers. At that time, the strategy followed by managers was to develop powerful techno-structures (engineering departments and HR specialists) to impose their monopoly over the work design through the de-socialization, de-skilling and dis-empowerment of the professionals. The intention was to put professionals under the control of organizations ruled by managers. Indeed, it is not an exaggeration to say that the managerial and bureaucratic rationalization of organizations was a war declared by managers against the professions and trades, and one that managers largely won. Throughout the 20th Century, professionals have been dramatically weakened inside organizations. In a sense, the paradox is that top managers are today the only true professionals that remain inside these modern complex organizations! In that specific context, the managerial call for the development of employee "professionalism" may create a great deal of frustration for employees who have clearly understood that it was not a wakeup call for a renewal of true professionalism based on strong professional identities but rather a call for ever more compliance with formal rules and expected behaviors as we described at the beginning of the chapter.

Are things really different today? Perhaps the war is over, and perhaps managers sincerely now call for empowered highly-skilled employees but the political ambiguities of "professionalization" remain. Although they may not be at war any longer, they are in a situation of "cold war" where external pressures coming from shareholders and regulators reinforce the ambiguities of the professionalization requirements. Ignoring them would give rise to misunderstandings about the deep causes of the success or failure of professionalization programs that consciously or unconsciously tend to lock professionals into a managerial/bureaucratic vision of professionalism.

As we stated earlier, the managerial/bureaucratic approach to professionalism promotes an ideal image of professionals that is inconsistent with their traditional model. The former is based on a particular quest for expertise with the following characteristics:

- The professional is an expert with perfect mastery of the technical and organizational aspects of his or her job, starting with the complete set of formal rules and procedures currently in use. He or she is supposed to be competent (i.e. able to act and make decisions properly in any kind of context) and makes no error and no faults. The "pro" is supposed to deliver relevant solutions simply by following the existing procedures. He or she is the actor that creates the organizational illusion that the reality matches the formal prescriptions of work.
- The "pro" is an autonomous and individual expert. On the one hand, expertise is supposed to be held by individuals, not directly by teams (viewed as sets of individuals); on the other hand, formal teams are considered to be the unique collective context within which the individual expertise unfolds and combines with others.
- The "pro" is an acculturated agent: his identity is melted into the identity of the organization he or she works in. They are supposed to accept and share internal best practices and the criteria in use in their company to assess the professional skills that ultimately shapes his or her professional identity.
- The "pro" is not a political actor. Instead, he or she is an autonomous expert... but without real power over the organization to which they belong. The "pro" does not question the objectives of the organization, he or she is loyal and the performance of the company prevails over his or her own interest without opportunistic behaviors such as free-riding.

Obviously, this bureaucratic representation is an illusion. The problem is that such an illusion is the basis upon which the organization decides who is a "pro" and who is not. It is also on this basis that professional development programs are designed and implemented.

The trade approach gives a very different image of professional development. First of all, there is not "one profession". Professions are numerous and are located in different communities, potentially competing with one another. Secondly, the identity of the professional is mainly shaped outside the organization (Barley 1996). It is rooted in professional communities ruled by professional guidelines and know-how and practices. Becoming a good practitioner, a "pro", refers to a process of socialization that entails becoming a member of a particular community. Some academics (Brown and Duguid 1991, 2001) speak about "communities of practice" that partly escape from the control of the formal hierarchical organization and promote a cross-functional logic. It is far from being limited to the unique judgment of managers and HR departments.

The problem is that the tensions between these two opposing representations of professionalism are not managed. We believe that safety professional development supposes to do so through reflexive and discursive organizational practices, according to High Reliability Organization theories.

11.2 Finding New Ways for Safety Professional Development: Managing the Tensions Through Reflexive and Discursive Organizational Practices

We assume that new ways for safety professional development have to be explored at the crossroad of the two opposite approaches of professionalism. This requires active management of the tensions between the two. Dynamic compromises have to be found regarding the roles and responsibilities of the professionals committed to safety, but also about their identity and power inside and outside their organization.

These analyses suggest that we should "bring work back in[to]" our understanding of organization and management (Barley and Kunda 2001) and put the focus on practices and actual working rules and discussions about work, especially in the field of safety management (de Terssac 2013).

We suggest that High Reliability Organizations theory (Roberts 1990; Weick and Roberts 1993) provides an interesting theoretical framework for the management of such tensions. HRO combine a very bureaucratic *organization* based on formal hierarchical structures, clear division of roles, formal rules, procedures and routines, with flexible "*organizing*" processes that come into play when the situation becomes highly complex and unexpected (Weick and Sutcliff 2007). HRO demonstrates that safety is rooted in the day-to-day activities of the practitioners who work in high risk industries. Safety is the final outcome of a continuous process of reflexivity that brings safety and daily practices into professional discussions. The key point is to keep the organization aware of safety problems and doubting about "what is going on" and "what should be done" (Weick and Sutcliff 2007).

We consider these reflexive and discursive organizational practices as levers for the management of some important tensions associated to the opposition between the managerial/bureaucratic approach and the professional/trade approach of professionalism. In the following sections, we present and discuss three of them.

11.2.1 Formal Safety Rules Versus Safety Embedded in Professional Practices, Knowledge and Debates

In the managerial/bureaucratic approach safety is supposed to be contained within formal rules. The legitimacy of the rules is rooted in the combination of de-contextualized scientific knowledge and hierarchical authority. In a professional perspective, safety is embedded in contextualized practices rooted in professional skills and expertise that shape the professional safety culture (Gherardi and Nicolini 2002).

In HRO, compromises between the two approaches can be elaborated in real-time action and then discussed and assessed after the facts in work discussions and work debate spaces (Rocha et al. 2015). In real-time action, people make sense of the problematic situations they must keep under control using cognitive resources provided by both formal and informal rules and norms. In such situations, the responsibility of professionals is to take initiatives and be empowered by doing so. Then, the decision "migrates" throughout the organization until it finds the right expertise, regardless to the hierarchical rank. A "self-designing" organization (Rochelin et al. 1987) emerges from the reflexive and heedful interactions people in the team develop to keep the situation under control (Weick and Roberts 1993). Once the action is over, people involved in the situation share their fresh experience and discuss how things were done, in a positive or negative assessment. These discussions confront and combine the formal hierarchical legitimacy with the professional one. They create the reflexive "experience" on which the professional builds up its expertise and becomes a "reflexive practitioner" (Schön 1983). In case of serious problems and doubts, other discussions can be organized in order to develop wider experience feedback learning loops. Managers should provide the resources to organize such reflexive practices. This managerial action supposes to design "work debate spaces" (Rocha et al. 2015) or "work discussion spaces" (Detchessahar et al. 2015).

In that sense, the safety professional is the reflexive practitioner who puts safety, safety practices and, even more widely, work into discussion and debates during action and after.

11.2.2   Training for Safety Versus Learning to Become
a Good Practitioner in Safety Industries

In the managerial/bureaucratic approach, safety training programs are aimed at learning safety rules, canonical practices and safety expected behaviors. In the professional approach, learning how to produce safety means learning how to become a reflexive practitioner (Brown and Duguid 1991). This is not just a question of knowledge and practice, it is a genuine process of socialization that organizes the entrance of the trainee into the group of professionals and modifies the identity of the trainee. It is also a political process that legitimates the knowledge and skills developed by the professional and gives him or her arguments for future professional debates.

11.2.3   Formal Teams Versus Professional Groups
and Communities

In the managerial/bureaucratic approach, professionals are supposed to work collectively in formal teams. These teams are determined by the way work has been divided in the organization. By contrast, professional groups refer to communities that don't always fit the structure of formal teams and departments. A professional group is a group that makes sense for its members. It is often described as a "community of practice" where members share the same practices and discuss them well beyond the organizational frontiers. Discussions are not necessarily consensual, they can be very challenging and take the form of debates and professional arenas where professionals compete and show off their skills. Internal relationships in such communities are both co-operative and competitive to produce shared professional norms and rules. It is the community within which the members find the resources to be a real professional. A professional group is also a political group that promotes the interests of the professional it represents.

In HRO, safety is based on auto-organized groups that emerge unexpectedly from collective action to quickly respond to a problematic situation. This is based on the "heedful interrelations" that practitioners develop among themselves to stay constantly aware of the situation, but also to mind and to care about colleagues who could potentially need help and support (Weick and Roberts 1993).

11.3   Conclusion: Discussion as a Fuel for the Professional
Development of Professionals and Managers

Our main conclusion is that "professionalization" of safety means reshaping the identity of the professionals working in high-risk industries. It is a real challenge for management because it requires finding acceptable compromises for both managers and practitioners. Communication and, more precisely, discussions and debates about safety appear to be the locus of "professionalization". Such discussions are not always spontaneous and need to be engineered and conducted by managers (Detchessahar and Journé 2011). This implies sharing knowledge, power and legitimacy inside and outside the organization. In that perspective, the notion of expertise cannot be reduced to simply an ability to know all the formal rules and to comply with them. A less bureaucratic approach of expertise and professionalism would include the ability to take some distance from the formal procedures and to discuss that. This raises the question of the collective dimension of the expertise as well as the social and political status of the "professionals" in high-risk industries.

Finally, we argue that the management of the tension between the managerial/bureaucratic approach and the professional/trade approach to safety professionalism does not imply having to "choose" between these two opposite representations. But rather, it means they should be combined in a way that strengthens the legitimacy of both of them. We assume that such a combination can be reached through discussions about safety practices that in turn question general safety principles and formal rules. The aim of the discussion is not to weaken the position of managers to the benefit of professionals. The outcome of the discussion should be the mutual empowerment of both managers and professionals. Discussion is a fuel for the professional development of both "professionals" and managers. In that sense, risky industries need strong (powerful and legitimate) professionals as well as strong (powerful and legitimate) managers to feed discussions about safety that aren't purely cognitively-based on the rational exchange of information, knowledge and opinions—but that are also based on organizational and political issues. This creates the responsibility for every participant to speak up, to listen and to draw the pragmatic consequences of the discussion. That is the reason why it is so difficult to organize such discussions. That is the reason why risky industries need to design and manage "work discussion spaces". And, that is the reason why training programs should be considered as privileged moments and areas for "discussion".

References

Barley S. (1996). Technicians in the Workplace: Ethnographic Evidence for Bringing Work into Organizational Studies. *Administrative Science Quarterly, 41*(3), 404–441.
Barley S. & Kunda G. (2001). Bringing Work Back In. *Organization Science, 12*(1), 76–95.
Berle A. & Means G. (1932). *The Modern Corporation and Private Property*. Harcourt, Brace & World inc.
Brown, J. S., & Duguid, P. (1991). Organizational learning and communities-of-practice: Toward a unified view of working, learning, and innovation. *Organization science, 2*(1), 40–57.
Brown, J. S., & Duguid, P. (2001). Knowledge and organization: A social-practice perspective. *Organization science, 12*(2), 198–213.
Detchessahar M. & Journé B. (2011). *The conduct of strategic Episodes: A communicational perspective*. 27th EGOS colloquium, July, Göteborg, Sweden.

Detchessahar M., Gentil S., Grevin A. & Journé B. (2015). *Organizing discussion on activity in complex project management: a key to articulate safety and performance in a high-risk industry.* 31st EGOS colloquium, July 2–4, Athens, Greece.

Fayol H. (1916). *Administration Industrielle et générale. Prévoyance, organisation, commandement, coordination, contrôle.* Paris: H. Dunod et E. Pinat (Eds.).

Gherardi S. & Nicolini D. (2002). Learning the Trade: A Culture of Safety in Practice. *Organization, 9*(2), 191–223.

Roberts C. (1990). Some Characteristics of One Type of High Reliability Organization. *Organization Science, 1*(2), 160–176.

Rocha R., Mollo V., Daniellou F. (2015). Work debate spaces: A tool for developing a participatory safety management. *Applied Ergonomics*, 46, 107–114.

Rochlin G, La Porte T. & Roberts K. (1987). The self-Designing High-Reliability Organization: Aircraft Carrier Flight Operations at Sea. *Naval War College Review, 40*(4), 76–90.

Schön D. (1983). *The reflective practitioner.* New York, Basic Books.

Taylor F. (1911). *The Principles of Scientific Management.* Harper & Brothers.

Terssac G. de (2013). De la sécurité affichée à la sécurité effective: l'invention de règles d'usage. *Annales des Mines- Gérer et Comprendre, 2013/1*(111), 25–35.

Weick K. & Roberts C. (1993). Collective Mind in Organizations: Heedful Interrelating on Flight Decks. *Administrative Science Quarterly, 38*(3), 357–381.

Weick K. & Sutcliffe K. (2007). *Managing the Unexpected: Resilient Performance in an Age of Uncertainty.* 2nd Edition, Jossey Bass, Wiley.

Chapter 12
Can Safety Training Contribute to Enhancing Safety?

Corinne Bieder

Abstract Training has always been an obvious response to any operational issue and safety issues are no exception. Further to an accident, training, and more specifically safety training, almost always forms part of the recommendations. More than that, safety training has always been considered by many as one of the major pillars for ensuring the safety of hazardous activities. This is the case in regulatory requirements as well as in many internal safety policies. Although this seems to make sense intuitively, intuition is not always of sound advice when it comes to safety. In reality, safety training conveys a number of implicit assumptions as to what contributes to making the operation of an organization safe. These assumptions, once made explicit, become debatable. However, unravelling them makes it possible to examine potential ways forward to reach beyond what seems to be the current safety training escalation dead-end.

Keywords Work practices · Regulatory requirements · Compliance · Safety performance

As provocative as it may sound, the discussions during the academic seminar led us to raise this fundamental question: can safety training contribute to enhancing safety?

The initial doubt was expressed by FonCSI's industrial partners, questioning the relevance of their safety training based on the perception or belief that their increasing investment in such training was no longer paying off as expected. However, in the light of the discussions, it appears that rather than asking how to deliver better or more efficient safety training, a more relevant question would be: are safety training courses an appropriate way to actually enhance safety?

This question emerges in reality from a deeper philosophical disconnect between two apparently opposite appreciations of safety:

C. Bieder (✉)
Ecole Nationale de l'Aviation Civile, Toulouse, France
e-mail: corinne.bieder@enac.fr

© The Author(s) 2018
C. Bieder et al. (eds.), *Beyond Safety Training*, Safety Management,
https://doi.org/10.1007/978-3-319-65527-7_12

111

- On the one hand, those who defend the concept of safety as fully embedded into work practices. This then translates into: doing your job well includes doing it safely where the idea of safety is built from experience following the theory of Aristotle. As such, safety can neither be thought out from scratch nor imposed by an external party. It is an intrinsic part of each singular situation and cannot be disconnected from its manifestations in the real world.
- On the other hand, those who defend the concept of safety as a distinct dimension of any job that can be thought out in a generic manner. It then translates into: working safely means comply with safety rules deriving from what would be the ideal Form of Safety in Plato's world of Ideas (Plato).

This disconnect has a number of implications that go beyond the skills and competences needed to operate safely. Indeed, it affects the very definition of what is considered to be a normal situation as opposed to an abnormal one and raises too wide a scope of questions for them all to be addressed in this chapter.

The academics invited to the workshop[1] confirmed this disconnect between these two understandings of safety in a significant number of big organizations with an often clear difference between the operational functions and the top management/support functions.

For those in operational roles, safety is seen as one dimension among all the others of their job (Cuvelier and Falzon 2011). In terms of training, it means that there is no such thing as a "safety training course" that would address the safety dimension in isolation from the rest of the job's requirements, environment or constraints.

Conversely, top managers or support functions envisage safety as an independent dimension of work, merely consisting of compliance with a number of exogenous requirements. Safety competencies can then be described and assessed regardless of the specific job and operational context. Safety can thus be taught in a generic manner independently from the rest. In other words, safety training courses are what is needed to enhance safety.

Indeed, once a training program is called or considered to be "safety training", it assumes to some extent that safety can be isolated from work practices, and even more so if the content of the safety training is generic to a number of industrial activities.

Although this approach is fully in line with a belief that fulfilling safety requirements is enough to ensure a safe performance, it is pointless in a belief that ensuring safety is about doing one's job well since the safety dimension cannot be dissociated from the other dimensions of the job (Bieder and Bourrier 2013).

From this common apparent deadlock, is there a way forward?

Before reflecting on possible avenues to explore, it is essential to return to the initial question from the industrial players: why is the increasing investment in safety training no longer paying off?

---

[1]The two-day international workshop mentioned in the preface, organized by FonCSI in November 2015 and highlight of the project that led to this book (editors' note).

We will not explore the actual safety benefits, or their absence, which often seem to rely on a perception or belief rather than on an actual measure, for this would require a whole paper. While existing safety trainings may seem to fail to produce results safety-wise, they nevertheless allow organizations to comply with regulatory requirements that call for safety training. Indeed, most regulatory authorities in hazardous activities are defenders of the "safety exists as such" belief. Or maybe should we say they used to be. Indeed, the evolution from an exclusively compliance-based approach to a more performance-based approach in some hazardous activities such as aviation may be initial evidence that the doubts expressed by FonCSI's industrial partners are shared, at least to a certain extent, by Authorities as well.

Up to now, the tension between the two apparent beliefs on safety, or safety models, has led to an increase in the effort in the direction of mandatory "safety trainings", often to the detriment of other initiatives focused on training courses that are better suited to enhancing safety. Yet, some initiatives in this latter direction were presented during the workshop with promising results.

Thus, the question becomes: is there a way of maximizing the resources dedicated to training (in a broad sense) that actually contribute to enhancing safety while complying with regulatory requirements? The most obvious answer would be through reconciling the two. Yet, regulatory requirements are developed to ensure that the minimum acceptable level of safety is ensured by all organizations of a given domain. Although they may stem from a safety model closer to one extreme than to the other, they are designed in a one-size-fits-all manner whereas each organization is unique.

Eventually, depending on the existing gap between the regulatory/oversight approach and the organization's maturity safety-wise, there may be different avenues to explore as ways forward:

- If the regulatory requirements and the oversight approach leave some leeway for interpretation, there may be a way to reconcile both aspects, actually enhancing safety and complying with regulatory requirements. By giving preference to the ultimate objective of the safety training rather than to a reductive interpretation of "acceptable means of compliance", revisiting the content, format… of these so-called "safety trainings" can be an opportunity to improve the actual safety performance.

  The introduction of mandatory CRM (Crew Resource Management) training in aviation following the most deadly accident in this domain in Tenerife is a very good illustration of how a similar requirement was translated into very different training courses by different airlines around the world. Interestingly enough, although it had a strong safety root, the requirement was not called "safety training".

  Regulatory requirements referred to a number of topics to be addressed during this training such as communication, leadership/followership, individual factors… Depending on the airline, CRM training courses ranged from strict basic 'teaching' on the various topics to more sophisticated and interactive sessions around these topics addressed through real-life examples. In other words, at one extreme, CRM trainings were generic theoretical lectures on communication and all the other required topics, disconnected from any realistic flight context, facilitated by human factors specialists with no aviation background or knowledge. At the other extreme, CRM training courses took the form of debates among professionals, initially pilots, based on anecdotes brought by participants taken from their own experience, facilitated by a pilot with additional human factors background or a human factors expert with additional flying background. Tricky situations, tips, procedure limitations and external pressures were discussed openly and shared among a group of professionals leading to a translation of regulatory topics into real work situations. Practices were discussed in the light of some theoretical inputs and a dialogue was engaged between professionals to cross-fertilize theory and practices (qualifying some theoretical aspects based on their limits in some singular experienced situations, as well as qualifying some practices that hadn't yet led to any unwanted events but could do so in slightly different contexts).

- If the regulatory requirements and the oversight approach provide strong incentives to develop training courses disconnected from work situations, isolating safety from the other dimensions, the key question becomes: can "safety training" resources be allocated differently, i.e. limiting the investments to the strict minimum necessary to comply with these requirements and investing further in something else than "more of the same" to actually enhance the safety performance? This would also mean dismissing the illusion that there is any safety benefit from mandatory "safety trainings" …

In his chapter, Vincent Boccara gave an illustration of a possible complementary approach through the creation of a discussion space around safety in work situations between defenders of the two apparently opposite beliefs on safety, to enable the debate as to how to actually ensure the safety of operations.

However, in the case of a significant disconnect between regulatory requirements and actual safety enhancement, a parallel avenue would be to explore whether there would be a way to revisit the regulatory framework and the safety training requirements—be it in their philosophy, focus, format…—so that they provide incentive to develop training that actually contributes to enhancing safety whatever the organization's initial maturity level in terms of safety?

Revisiting the regulatory and oversight framework requires a holistic approach in terms of all the dimensions that are impacted by switching from a compliance-based approach to a performance-based approach to safety. While the issues of empowerment, accountability, control, expertise, etc., were discussed extensively in relation to how safety is managed within an organization, there is a mirror situation at the level of the regulator or between the regulator and the organizations it oversees that needs to be considered in its complexity, keeping in mind the additional challenge of not belonging to the same organization or sharing the same goals… What is the actual work practice of a regulator and how does safety as a situated work practice translate in this environment are important preliminary questions to analyse.

In this framework, working on an adaptation of Boccara's approach, which seeks to stimulate debate between operating organizations and their regulator with regard to work situations (to be defined or extended) could possibly contribute to making regulatory requirements evolve, at least in their flexibility.

The shift from a compliance-based regulation approach to a performance-based one initiated in some hazardous activities (ICAO 2013) should allow the regulatory and oversight approaches, including the "safety training" requirements, to be significantly revisited. However, if the regulatory approach is to develop in this way, this will also involve evolutions in a number of areas that reach far beyond the wording of the regulatory requirements themselves, whether they refer to external requirements developed by the institutional external Regulator or by the internal relays of the Regulator's exogenous requirements (e.g. Quality department...). How to make the practices of "rule-makers" (both external and internal) evolve in an appropriate direction to support a performance-based approach to safety is not an easy question. Part of the answer is probably based on "training" in a very broad sense, but maybe not on "safety training".

References

Bieder, C. & Bourrier, M. (2013), *Trapping Safety into Rules,* Ashgate.
Cuvelier, L. & Falzon, P. (2011). Coping with uncertainty. Resilient decisions in anaesthesia. In E. Hollnagel, J. Pariès, D.D. Woods and J. Wreathall (Eds.), *Resilience Engineering in Practice: A guidebook*, Ashgate studies in resilience engineering. Ashgate.
ICAO (2013). *Annex 19, Safety Management.* Montreal, Canada: International Civil Aviation Organization.
Plato, Phaedo. http://classics.mit.edu/Plato/phaedo.html

Chapter 13
Training Design Oriented by Works Analysis

Vincent Boccara

Abstract This chapter presents an approach to training design oriented by a holistic real-world works analysis based on several works of research. This approach proposed to design training in order to make people able to deal with real-world work situations, rather than only to know and apply exogenous standards. Two main axes of progress in the design of vocational training are identified and could develop into guidelines in order to train people to deal with work situations. (1) The approach requires project management in order to use participatory methods, including end users (trainers and trainee) and integrate a works analysis. (2) The approach needs to move from classical teaching-learning methods to "active" methods, which often imply transformation of both the trainer and the trainee's activity. Examples from previous research in training design are presented to illustrate the argument.

Keywords Real work situation · Vocational training · Participation

13.1 Introduction

We share the three conclusions made in the initial call concerning safety in industry (Foncsi 2015):

1. an isolation of safety from the other dimensions of work,
2. a disembodiment of work situations, and
3. the view that training is designed and implemented by stakeholders who are guided in particular by concepts of accountability and compliance with exogenous standards.

However, these three points do not apply only to the topic of safety. Systems of professionalization frequently focus on technical and regulatory contents, isolate

V. Boccara (✉)
Université Paris-Sud, Paris, France
e-mail: boccara@limsi.fr

© The Author(s) 2018
C. Bieder et al. (eds.), *Beyond Safety Training*, Safety Management,
https://doi.org/10.1007/978-3-319-65527-7_13

some specific dimensions of work—such as safety—from all others, and are designed based on a frame of reference. Let us also note here that these frames of reference are often originally designed for reasons other than training, such as regulatory purposes for example. It seems, then, that the topic of the call questions the very purpose of professionalization systems.

From this perspective, the present contribution aims to present an approach to design formal courses of professionalization based on real-world work activities. It is based on two main principles that will be exposed here:

1. professionalization of workers is a living and dynamic process;
2. formal training courses must be designed that are oriented by the analysis of real-world work situations. This approach might go some way to answer the question posed in the call: *"Is professionalization a safety issue… or the other way around?"*. In this chapter, we will briefly outline some noteworthy points concerning professionalization and the theoretical background of the proposed approach.

We will then address several guidelines based on research in training design in high-risk domains.

13.2   Professionalization: A Long-Term Living and Dynamic Process

"Professionalization" could refer to the process by which a person acquires the acceptable competencies and qualifications recognized by a professional body, or more broadly by a professional institution. It is therefore an individual and a social process that creates a dividing line between qualified workers, unqualified workers, and amateurs and is thus related to a frame of reference of required competencies and a formal system of qualification.

In these terms, the individual process of professionalization refers, for the workers, to the development of their competencies during their career. "Professionalization" is then a living and dynamic process oriented by the development of the ability to cope with work situations. It should be viewed as an on-going, long-term process related to the career path rather than a piecemeal process. One challenge is therefore how to design formal training courses in order to efficiently support the professionalization of workers.

All kinds of learning and training situations could refer to "professionalization". These include formal training systems as well as informal, on-the-job learning. Professionalization systems should aim to make individuals able to cope with the work situations with which they are confronted on a daily basis, in all of their complexity. Seen in this way, the challenge of vocational training systems would be to foster the development of vocational competencies that can be effective in work situations—that is, competencies that integrate the issues of production, safety and quality, as well as health. In order to achieve this, the design of a formal professionalization system must be guided by real-world work situations. This refers to how one can transfer work situations to training situations, in order to support the development of skills, whilst ensuring that they also favour the transfer of these skills in future work situations. These dialectics invite us to think in terms of vocational training courses rather than in terms of mere sequences of training sessions that are decoupled from an inscription in people's vocational history. In the same way, it suggests a need to think about the continuity and breakthroughs between the learning potential of work situations and training situations, with the goal of promoting the articulation between the two.

Hence, the goal of professionalization systems becomes knowing how to construct the potential for development that lies in situations of training (Mayen 1999) and/or of work (Falzon 2014), in order to help individuals develop the skills that are sought after by the organization. And in the case of safety-related questions, it would be advisable to include the issues that are inherent to work situations without separating them from other dimensions: the tasks that are to be completed, the questions of performance, collective dimensions, health, etc.

13.3 An Activity-Based Approach to Design Vocational Training Situations

We therefore propose a holistic approach to the design and assessment of vocational training courses, stemming from the contribution of ergonomics (Falzon 2014) and vocational didactics (Pastré 2011), both of which refer to the concept of work activity (Daniellou and Rabardel 2005).This alternative approach is based on a set of principles regarding human activity and its conditions of elaboration (Daniellou and Rabardel 2005) that are very useful for designing situations that foster learning and the development of vocational competencies. These principles include:

1. Activity is situated, in two senses. Individuals act depending on the situations they have to cope with, which are variable, evolve in time, and require changes in the activity; and activity is also marked by the period and culture in which it takes place;
2. Activity is finalized, oriented by goals which are partly specific to the individual carrying it out. His/her goals may be different from the goals that have been prescribed by exogenous stakeholders;
3. Activity is integrative, it emerges at the intersection of the individual's own features (his/her aims, personal history, knowledge, skills, etc.) and the features of situations in which he/she acts (the goals that must be met, the material means provided, the work environment, etc.);
4. The activity developed by a specific individual in a given situation is unique: it inherits elements from his/her past. It is constantly revised and renewed.

In this perspective, learning and the development of vocational competencies are both a process and a product/result of activity. More specifically, they are a result of the constructive dimension of activity, which is the dimension oriented towards the elaboration of resources for oneself in future situations (Rabardel 1995/2002). Accordingly, it is then irrelevant to directly change external determinants of activity such as knowledge or competencies independently of situations. In contrast, we can act through the mediation of situations. That is why the notion of "situation" becomes a key element in the design of vocational training. The design of situations that entail a potential for development (Mayen 1999) leads us to question two dialectic processes: didactical transposition from the work situation to the training situation and transfer from the training situation to the work situation (Samurçay and Rogalski 1998).

This implies a shift in focus with respect to classical approaches of pedagogical engineering in adult training (Carré and Caspar 2011), which are widely used in the professional world. Following this classical approach, the contents of a training programme are derived from a frame of reference—related for example to regulations, to a specific trade, to technology, to skills, etc.—which people must then be trained to. In other words, this form of professionalization is based on *prescribed* work, which is necessary—but not sufficient—to cope with the realities of work. We must also move forward from a view of professionalization that is focused on the concept of knowledge, since the goal becomes being able to cope with work situations. We must, once again, keep our distance from a modular view of learning, in order to move towards a vision centred on the concept of training courses. This includes the temporal aspect of learning in the context of training courses. Lastly, we have to move away from a vision centred on regulations and/or prescriptions in order to integrate real-world work and the debate between real work and prescribed work.

13.4   Guidelines for Designing Vocational Training from Research in the Field

This approach is the result of an ongoing research programme concerning the development of vocational skills and the design of training systems, tools and situations based on work analysis. This programme hinges on several works of research in domains that involve risks, such as automobile driving (Boccara 2011; Boccara et al. 2014, 2015), aeronautics (Boccara and Delgoulet 2013, 2015; Delgoulet et al. 2015), work in a civil nuclear power plant (Fucks and Boccara 2014; Couix et al. 2015) or medicine in theatres of war (Delmas et al. 2015, ANR Project VICTEAMS).

Based on this research work, two main axes of progress in the design of vocational training have been identified and could become the basis for guidelines to train people to deal with work situations.

1. The approach requires project management in order to use participatory methods, including with end users (trainers and trainees) and integrate an analysis of works.
2. It needs to move from classical teaching-learning methods to "active" ones that often imply transformations of both the trainer and trainee activity.

13.4.1 Building a Participative Approach to Training Design Oriented by Works Analysis

The approach presented requires the inclusion of "works analysis"[1] in the management of a training design project (Boccara and Delgoulet 2013). The scope of the analysis is not just the task or job that is the focus of the training, as is the case in the classical approach of training design (IAEA 1996). Instead, as a minimum, it involves analysing the work of production operators, of trainers and of trainees. The work analysis consists in

> the global approach, where the activity analysis is integrated into an analysis of the economic, technical and social factors with which the operator is faced, and an analysis of the effects of the company's operations on the population in question and of economic efficiency. (Daniellou 1996, p. 185, our translation)

However, this "works analysis" must be conducted at the crossroads of approaches defended in ergonomics and vocational didactics (Boccara and Delgoulet 2015), because it is guided by the design of training/learning situations. This kind of works analysis aims to highlight the real-world work situation in production and in training, in order to identify the multiple horizons of the training situations to be defined: "training for what?", "training how?", "what device(s)?", "for what purposes?" (Olry and Vidal-Gomel 2011), considering working conditions for training to be learning conditions for trainees (Chatigny and Vézina 2008). More specifically, this works analysis method provides the identification and analysis of characteristic situations of action (Daniellou 2004) in order to recommend training objectives (Olry and Vidal-Gomel 2011) by formalising baseline professional knowledge (Samurçay and Rabardel 2004) as well as didactic transpositions (Samurçay and Rogalski 1998). This refers to identifying the differences and similarities between these two types of situation from the point of view of the activities deployed, those which cannot be deployed, and those which it would be recommended to deploy in order to improve trainee learning and the development of their activity. The analysis of these differences might direct training course design and anticipate the modifications of the activity of trainers and trainees, and its conditions of realization. For example, we proposed the SITUAATING method

---

[1]We deliberately use the term "works analysis" with a "s" at the end of "work". The reader can find a detailed explanation of this conceptual choice in Boccara and Delgoulet (2015).

that could be integrated in a design training management project based on research in civil nuclear power plants domain (Couix et al. 2015). SITUAATING is a proposal for moving from a classical approach in design training—like *Structured Approach to Training* (IAEA 1996) in the nuclear domain—to an approach where training situations are designed from works analysis.

The project management must also question the strategic orientations of professionalization throughout the training design, e.g. those related to safety: do organizations wish to make their stakeholders able to cope with hazardous situations? Or are they restricting themselves to just enforcing and complying with the rules? Such an approach thus means safety issues must be documentated, taking into account the diversity of work situations from the point of view of the organization and of its stakeholders, in the early stages of the training design process. Analysing the policies and structures surrounding professionalization, as well as the involvement of stakeholders (in particular managers and trainers) must be then an integral part of project management, as they are necessary conditions for driving evolutions in the content and orientations of training. For example, in the NIKITA project,[2] a phase of the analysis was dedicated to involving the strategic actors at the partner's site of the project in order to organize their participation in the project (Boccara & Delgoulet, 2013). Then, we progressively integrated several actors: trainers, teachers, trainees, prevention department (occupational risk management staff, occupational doctor), internal ergonomist, journeymen, team managers, line managers and the CHSCT.[3] This phase helped to resize the scientific project through a re-examination of its "intentions" (Barcellini et al. 2014), not just by putting into perspective the knowledge developed in relation to the actual work and technological possibilities, but also in terms of the relevance, from a work standpoint, of the scientific options that had been validated up to that point. It led to a shift from the initial objective of using virtual reality to teach "professional gestures" within a course built around a future Virtual Training Environment (VTE) which would take full charge of the trainee, to a project, within a broader training context, to design a VTE which would help to develop the cognitive organization of the action when performing assembly tasks. It also highlighted the crucial role of the trainer in the learning which meant designing the VTE as a working tool for trainers as well as a learning tool for trainees. Furthermore, changing the intentions of the project from professional gestures to training in the cognitive organization of action impacted directly on the content and situation of training, and consequently the tool that we had to design.

---

[2]The "Natural Interactions, Knowledge, Immersive system for Training in Aeronautics" (NIKITA) research project, funded by the Agence National pour la Recherche (ANR) and coordinated by Domitile Lourdeaux from the Heudiasyc laboratory at the Université Technologique de Compiègne (http://www.emissive.fr/nikita/). This project aimed to build a Virtual Training Environment (VTE) for aeronautical assemblers.

[3]The committee of hygiene, risks and work conditions.

13.4.2 How to Support Trainer-Trainee Work Activity in Order to Improve Professionalization?

If the end goal of professionalization in the domain of safety becomes "making workers able" to manage risks (together) in an integrated fashion, within situated productive activities, there will be an impact on the trainer's activity and on his/her role within the institution/organization. Both these points then deserve to be questioned using work analysis-based approach. Promoting such a goal for professionalization implies designing, implementing and "capitalizing" on forms of pedagogy and didactics where the trainees are actors within the situations with which they are faced. The term "situation" here incorporates the dimension of the organization of work as a social dynamic construct.

Within these situations, they must act and debate in order to learn and develop competencies. Furthermore, these training situations may serve as a medium to stage and to play out the issues surrounding the activity, the tasks, the variability, the problems and controversies that lie in real-world work. This implies, in particular, that trainers should be able to identify and "didactize" these objects belonging to situations of work, in order to turn them into objects of learning. The goal is also for trainers to have the power to "put up for debate" and "organize the debate" about rules and practices—whether these be prescribed or developed over time through the trade and professional customs with and between the trainees. This requires new pedagogical methods and tools to support the activity of trainers and trainees. The method of simulating works activities could be useful because it offers many possibilities and combinations to transpose the characteristics of real-world work activities: map, room training, verbal simulation, numeric simulation, full-scale simulation, etc. For example, Barcellini et al. (2014) proposed a method to simulate work organization for the design of work situations. This method could be transposed in the domain of training to simulate the organization of work situations, particularly to train mentors and managers in pairs or in cross-training with their crews. In the same perspective, work-based gaming tools could also be used to learn, discuss and debate rules and work process knowledge in order to make trainees able to cope with normal, daily and degraded situations.

More particularly, we built an ad hoc scenario-based gaming tool like this as a tool intended for trainers and trainees in cross-professional initial training to manage classical and radiological risks for workers in civil nuclear power plants in France (Fucks and Boccara 2014). The didactic tool is based on the real-world work process in order to create opportunities for trainees to experiment and discuss decision-making processes according to different work situations. The tool was designed to replicate different scenarios of work situations. Trainers could also adapt and increase the level of difficulty in terms of situation management and conflict resolution with and between the prescribed rules. Hence, this tool made it possible to discuss real work situations during the training, including the construction of the problem, as well as the way or ways it should be dealt with.

Following this perspective, the role of the trainer and of the trainees had to be analysed. The role of the trainer should not only be that of a guide, a coordinator or animator of contents (frames of reference, knowledge, rules, procedure, case studies, etc.) produced by others without them. Neither should the trainer be a "midwife" of knowledge without having any conceptual competence or experience in the trades and work situations they are training people in. Conversely, the role of the trainer cannot be solely to be an expert of the trade and of its related work situations, without having any pedagogical or didactic resources that make it possible to learn and to accompany workers in training in the construction of the desired competencies and skills. The goal here is to construct new trade-offs between the professional competencies of the trainers in the trades involved—viewed as objects of training—and the design, running, and evaluation of training programmes that are guided by and intended for work. In other words, this perspective leads us to question more broadly the work of trainers in sociotechnical production systems: Who can become a trainer? How does one become a trainer? What does the organization expect of the trainer? What is/are the possible career path(s) as a trainer? What is the future for trainers in organizations? What is/are the courses available to trainers for career development? These questions need to be answered and remain on the table throughout the training design process, because they deal with factors that may impact learning.

13.5 Conclusion

By way of a conclusion, we presented in this paper an approach to training design on the basis of works analysis in order to train people to deal with real-world work situations. We highlighted the fact that achieving this goal requires project management to take participatory methods into account, involving several actors of the company (trainers, trainees, managers, technical experts, etc.) and integrating an analysis of works both in training and in production. Training design needs to move from classical teaching-learning methods to "active" ones that often imply transformations of the activity of both the trainer and of the trainees. Thus, the dialectics between situations of work and training also invite us to think about the notion of *course of professionalization*, rather than about sequences of "moments of professionalization" to be inscribed in the career paths of the individuals involved. This suggests a need to think about continuity and/or breakthroughs in work situations and training situations, in order to foster connections between the two. Hence, this view questions the relationship between production and training in sociotechnical systems, in a more global and strategic manner.

Returning to the NIKITA project, the VTE was anticipated as an innovative tool in a training programme for temporary workers. The training system was based on a two-month period of initial training, followed by a period of several months of monitored work at the workstation. The workers were then considered to be autonomous. The workers could also have specific complementary training according to the specific features of their workstation. As a reminder, the aim of the project was to build a virtual training environment. Designing a new training tool thus involved analysing the existing training system and tools in order to identify where it could be integrated, its objectives, and how it could lead to additional benefits for trainees as well as for trainers. In other words, it is essential to think about the complementarity and the compatibility of the different "components" of the training system (sequence, module, situation, etc.) from the early stage of training design.

If we extend this idea, this orientation requires data from human resources to be organized in terms of career paths, going beyond merely tracing job changes over the years. It therefore questions the managerial processes involved in identifying training-related needs and their evolution over time, in service of a professional "trajectory". In other words, this requires longitudinal—rather than yearly—management practices, and questions the managerial and human resources processes of companies.

References

Barcellini, F., Van Belleghem, L. et Daniellou, F. (2014). Design projects as opportunities for the development of activities. In P. Falzon (Ed.), *Constructive ergonomics*. USA: Taylor and Francis.

Boccara, V. (2011). Développement des compétences en situation de tutelle au cours de la formation à la conduite automobile. Apports croisés de la psychologie ergonomique et de la psychologie sociale. Doctoral dissertation, Université Paris 8, St-Denis.

Boccara, V., & Delgoulet, C. (2013). Articuler les démarches d'analyse du travail en ergonomie et en didactique professionnelle pour la conception d'un EVAH. *Journées Scientifiques de Nantes*, 5–7 June 2013.

Boccara, V. & Delgoulet, C. (2015). Works analysis in training design. *Activités* [Online], 12–2|, mis en ligne le 15 octobre 2015, retrieved 10 April 2017. URL: http://activites.revues.org/1109; DOI: 10.4000/activites.1109.

Boccara, V., Vidal-Gomel, C., Rogalski, J., & Delhomme, P. (2014). Concevoir des référentiels comme des outils pour les formateurs? Réflexions à partir de la formation initiale à la conduite automobile. In B. Prot (Ed.), *Référentiel, Compétences, Développement* (pp. 119–132). Toulouse: Octarès.

Boccara, V., Vidal-Gomel, C., Rogalski, J., & Delhomme, P. (2015). A longitudinal study of driving instructor guidance from an activity-oriented perspective. *Applied ergonomics, 46, 21–29.*

Carré, P. & Caspar, P. (2011). *Traité des sciences et techniques de la formation* (2^d edition). Paris: Dunod.

Chatigny, C., & Vezina, N. (2008). L'analyse ergonomique de l'activité de travail: un outil pour développer les dispositifs de formation et d'enseignement. In Y. Lenoir (Ed.), *Didactique professionnelle et didactiques disciplinaires en débat* (pp. 127–159). Toulouse: Octarès.

Couix, S., Boccara, V. & Fucks, I. (2015). Training design for a not yet existing activity: the case of Remote Monitoring System for Risk Prevention (RMSRP) operator in French Nuclear Power Plants. *19th World Congress Ergonomics*, 9–14 August. Melbourne: Australia.

Daniellou, F. (1996). *L'ergonomie en quête de ses principes*. Toulouse: Octarès.

Daniellou, F. (2004). L'ergonomie dans la conduite de projets de conception de systèmes de travail. In P. Falzon (Ed.), *Ergonomie* (pp. 359–373). Paris: PUF.

Daniellou F., & Rabardel P. (2005). Activity-oriented approaches to ergonomics: Some traditions and communities. *Theoretical Issues in Ergonomics Science, 6* (5), 353–357.

Delgoulet, C., Boccara, V., Carpentier, K., & Lourdeaux, D. (2015). Designing a virtual environment for professional training from an activity framework. Dialog between ergonomists and computer scientists. *19th World Congress Ergonomics*, August 9–14. Melbourne: Australia.

Delmas, R., Boccara, V., & Darses, F. (2015). Analyse de la prise de décision collective en situation de crise pour la conception d'environnement virtuel de formation. *8ème colloque EPIQUE*, 8–10 July, Aix-Marseille.

Falzon, P. (2014). *Constructive ergonomics*. USA: Taylor and Francis.

Foncsi (2015). Is professionalization a safety issue…or the other way around? Call for papers. https://www.foncsi.org/fr/media/foncsi-as-2015-professionalization-safety-issue.pdf

Fucks, I. & Boccara, V. (2014). Les défis de l'intégration de l'expérience professionnelle dans des formations « multi-métiers » ? 3ème congrès international de didactique professionnelle « Conception et formation ». Caen, 28–29 October 2014.

IAEA (1996). Nuclear power plant personnel training and its evaluation: a guidebook. *Technical Reports series* N°380. Vienna: IAEA.

Mayen, P. (1999). Des situations potentielles de développement. *Éducation Permanente, 139*, 65–86.

Olry, P., & Vidal-Gomel, C. (2011). Conception de formation professionnelle continue: tensions croisées et apports de l'ergonomie, de la didactique professionnelle et des pratiques d'ingénierie. *Activités, 8*(2), 115–149. http://www.activites.org/v8n2/v8n2.pd.

Pastré, P. (2011). *La didactique professionnelle*. Paris: PUF.

Rabardel, P. (1995/2002). People and technology: a cognitive approach to contemporary instruments. Université Paris 8, pp. 188. https://hal-univ-paris8.archives-ouvertes.fr/file/index/docid/1020705/filename/people_and_technology.pdf.

Samurçay, R., & Rabardel, P. (2004). Modèles pour l'analyse de l'activité et des compétences, propositions. In R. Samurçay & P. Pastré (Eds.), *Recherches en didactique professionnelle* (pp. 163–180). Toulouse: Octarès.

Samurçay, R., & Rogalski, J. (1998). Exploitation didactique des situations de simulation. *Le Travail Humain, 61*(4), 333–359.

Chapter 14
Safety and Behaviour Change

Paul M. Chadwick

Abstract Promoting industrial safety is a complex field requiring collaboration between academia and industry across a range of professional and academic disciplines. Whilst human factors are recognized as being key modifiable determinants of risk across all professional groups and disciplines the variety and type of theories, methodologies and practices can make it difficult to identify commonalities and integrate findings into a conceptually coherent framework for research and intervention. The science of behaviour change offers possibilities for integrating cross-disciplinary understandings of the contributions of human behaviour to industrial safety through the use of models and frameworks like the Behaviour Change Wheel (BCW). This chapter describes the principles and processes involved in designing behaviour change interventions using the BCW illustrating this with examples drawn specifically from the industrial safety sector. The potential applications of the approach in the areas of workforce development and research are highlighted.

Keywords Capability · Opportunity · Motivation

14.1 Introduction

Whilst increasing amounts of resource are ploughed into initiatives to improve industrial safety there appears not to be a corresponding return on this investment as manifest by outcomes such as a reduced frequency of accidents or major hazards. Human behaviour has been identified as a major modifiable determinant of exposure to risks and hazards and is widely agreed to be a legitimate target of interventions to improve safety. Whilst numerous theories and frameworks have been used to understand and intervene with the behavioural determinants of risk the field

P.M. Chadwick (✉)
Centre for Behaviour Change, University College, London, UK
e-mail: p.chadwick@ucl.ac.uk

© The Author(s) 2018
C. Bieder et al. (eds.), *Beyond Safety Training*, Safety Management,
https://doi.org/10.1007/978-3-319-65527-7_14

suffers from a proliferation of models which are difficult to compare and contrast. This limits the accumulation of a coherent body of knowledge and expertise about what works, in what situations, for what problems, both within and across sectors. Furthermore, practices in the field may or may not reflect what is known scientifically, and the rise of the 'safety industry' and 'safety professional' means that the theories and techniques behind methodologies may be obscured by commercial interests.

This chapter examines the potential contributions of the emerging science of behaviour change to the field of industrial safety. It will outline the principles of understanding and changing behaviour using the Behaviour Change Wheel (BCW); a theory and evidence-based framework for designing behaviour change interventions that is gaining traction in cross-disciplinary research (Michie et al. 2014).

14.2 The Emerging Science of Behaviour Change and the Behaviour Change Wheel

The study of behaviour change has its roots in experimental psychology. As such, the tools of the scientific method—theory, hypotheses, experimentation, evaluation —are at the heart of the approach. Whilst there has been a great deal of empirical research into behaviour change across a range of sectors—including industrial safety—the field has lacked a unifying framework by which findings from studies employing different theories and methodologies can be integrated. In many areas, this has resulted in a fragmented research literature that can be difficult to pull into a coherent body of knowledge for the purpose of designing interventions.

One recent approach to reducing this muddle and providing coherence is the Behaviour Change Wheel (Michie et al. 2014). The Behaviour Change Wheel is a synthesis of 19 frameworks of behaviour change identified across a range of behavioural and social sciences (Michie et al. 2011). The BCW consists of three layers (Fig. 14.1).

The approach begins at the hub of the wheel where the sources of the behaviour that could prove fruitful targets for intervention are identified (i.e. a behavioural analysis). A simple model of behaviour, COM-B, is used to conduct the behavioural analysis. COM-B is an acronym for 'Capability' (physical and psychological), 'Opportunity' (physical and social) and 'Motivation' (automatic and reflective), conceptualised as the three essential conditions for behaviour. Surrounding the COM-B model is a layer of nine intervention functions that can be used to address deficits in one or more of capability, opportunity or motivation. These intervention functions can then be linked to the behaviour change techniques (BCT's; i.e. the active components of an intervention) described in published taxonomies of BCTs (Abraham and Michie. 2008; Michie et al. 2013). Finally, the outer layer, the rim of the wheel, identifies seven types of policy that one can use to deliver the intervention functions.

Fig. 14.1 The behaviour change wheel (Michie et al. 2011)

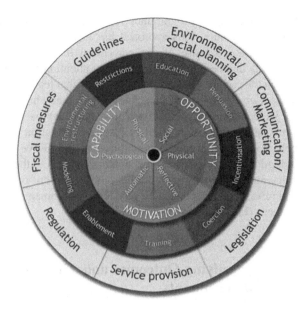

In order to translate general intervention functions identified using the Behaviour Change Wheel into a practical intervention for the given context, relevant behaviour change techniques are subsequently identified (for more information about behaviour change techniques, see Michie and Johnston 2011). Further consideration as to how the behaviour change techniques will be delivered within organisations or communities is taken into account by considering criteria such as the affordability, practicality, effectiveness and cost-effectiveness, acceptability, side-effects/safety, and equity of the intervention (i.e. the APEASE criteria; Michie et al. 2014). The BCW is designed to be a pragmatic framework that can be used to systematically design and support the evaluation of behaviour change interventions (see Michie et al. (2014) for more details).

14.3 Behaviour Change Versus Behavioural Safety Approaches

The behaviour change approach described in this paper should be differentiated from the 'safe behaviour,' 'behaviour modification' or 'behavioural safety' approaches described by Hopkins (2006). Whilst behaviour and how to change it is at the heart of both approaches, 'behavioural safety' programmes are more narrow in focus and deal primarily with downstream causes of accidents. Theorists have criticized the behavioural safety approaches for falling foul of the 'fallacy of monocausality', which is the idea that there is often a single root cause—in this case, behaviour—of an event. Since human factors are often implicated at some point during the causal chain of events leading up to an industrial accident, behavioural safety programmes focus on understanding and modifying the associated behaviours. This approach has been heavily criticized by unions and academics alike because of the perception that it unfairly blame workers for the accidents that befall them, misdirecting attention from other factors that may play a role in the complex causal chain of events that leads up to an accident. In contrast, a *behaviour change* approach acknowledges that whilst unsafe behaviour may trigger an accident, the behaviour itself may be better viewed as something requiring an explanation rather than itself being the explanation (Hopkins 2006). As such, the behaviour change approach outlined in the BCW acknowledges that changing individuals' behaviour requires making changes to the contextual influences on that behaviour. This may include the way people are organized, managed, motivated and rewarded, as well as their physical environment and the tools that are available to them (Fleming and Lardner 2002).

14.4   Specifying Outcomes and Their Behavioural Determinants

The starting point for any attempt to change behaviour is to determine what it is that needs to be different (the outcome) and then identify the behavioural determinants of that outcome. For example, if a company wishes to have fewer on-site accidents then it would need to specify the specific types of accidents (e.g. fewer machine tool injuries) and the behaviours that are related to this (e.g. using tools correctly, wearing personal protective equipment). Specificity is important since all behaviour is context dependent. The behavioural determinants of one type of machine tool injury may or may not translate to a different machine because different machines require different types of complex motor movements to operate them. Similarly, the behavioural determinants of a particular type of machine tool injury on the same machine may be different between sites because the context in which the machine and the worker interact may be different. There may be commonalities across contexts but one should always be open to the possibility that the behavioural determinants of outcomes could vary in important ways between them.

14.5   Behaviour Change, Safety-I and Safety-II

Accidents caused by deviations to established protocols for routine tasks are different to accidents caused by workers' behavioural responses to unexpected events that have never happened before, or cannot be foreseen. Interventions to prevent accidents arising from the former are referred to as Safety-I approaches, whilst interventions directed towards limiting the damage caused by the unexpected and unforeseen are termed Safety-II approaches (Hollnagel et al. 2015). A behaviour change approach can be applied within both approaches since they share an emphasis on understanding behaviour in context. For example, being able to specify the behavioural determinants of effective responses to unexpected events can inform how organisations train individuals to respond to similar circumstances. In both cases the behaviour change approach is based on an understanding of how people actually carry out their work (work as done) as opposed to how they are supposed to do it (work as imagined). Interventions developed with the BCW always start from examining the behaviour in context and in many cases this requires changing the context to enable the behaviour rather than changing behaviour to fit the context. Within the industrial safety sector, an intervention based on behaviour change principles may well involve restructuring the nature of work itself in order to bring it in line with the known limits and constraints on human performance, rather than attempt to modify human performance in order to bring it in line with unattainable production targets.

Whilst behaviour change principles are amenable to looking at safety through both Safety-I and Safety-II perspectives, the BCW may at first seem more aligned with Safety-I approaches on the basis that the primary unit of analysis is behaviour. Nevertheless, by understanding behaviour as context-dependent, and broadening the definition of context to include the organisational, cultural and linguistic determinants on behaviour, models such as the Behaviour Change Wheel have the potential to be applied within Safety-II frameworks. Safety-I looks at why things go wrong and tries to identify and eliminate the causes of error, whereas Safety-II looks at why things go right and tries to ensure that they happen again, often by promoting organisational resilience. In both cases a focus upon describing the situation in behavioural terms and identifying the determinants of what goes wrong, or what goes right, is likely to generate helpful new insights for targeted interventions.

14.6 Specifying What Needs to Change—Behavioural Diagnosis

Once the behavioural targets have been identified the next step in designing a behaviour change intervention is to identify the determinants of the behaviour(s). The Behaviour Change Wheel uses a simple model of behaviour, COM-B (Michie et al. 2011) as a framework to categorise the various influences on the behaviour. COM-B is an acronym for the three essential conditions for behaviour; 'Capability' (physical and psychological), 'Opportunity' (physical and social) and 'Motivation' (automatic and reflective).

14.6.1 Capability

Capability refers to an individual's ability to carry out the required behaviour. The model distinguishes between physical and psychological capability, the former being the physical skills, strengths and stamina to behave in a certain way, for example, being fit enough to continue handling well at the end of a long shift, the latter being the necessary mental process required to carry out a behaviour, such as being able to reliably make the right decision in the face of a barrage of complex information.

14.6.2 Opportunity

Opportunity refers to those influences on behaviour that are largely external to the individual and are found in the physical and social environment (e.g. productivity targets). Physical opportunity refers to the time, resources, locations and cues that trigger behaviour or enable it. Social opportunity refers to the interpersonal influences on behaviour such as behavioural norms, peer influences, role models, as well as broader aspects such as the linguistic and cultural concepts that shape behaviour and its expression (e.g. safety and management culture).

14.6.3 Motivation

Motivation refers to all those processes, conscious and unconscious, that energise and direct behaviour. The model distinguishes between reflective and automatic processes. Reflective motivation refers to the conscious plans, beliefs, desires and intentions that influence behaviour, such as the specific intentions to behave in ways that are consistent with an individual's beliefs about their identity. Automatic motivation refers to the largely unconscious influences that shape behaviour, such as emotional reactions (e.g. the experience of guilt or shame if found to be doing an unsafe behaviour), impulses, inhibitions and drive states such as hunger and thirst, and habits (e.g. reaching for a lever that is typically on a certain side).

14.6.4 Pulling Together the Behavioural Diagnosis

Capability, opportunity and motivation all act to influence the expression of behaviour in a reiterative way. For example, an organisational culture characterized by high levels of trust between workers and management (social opportunity) may lead to greater engagement with initiatives to improve behavioural safety (reflective motivation) leading to workers who have the acquired physical and psychological skills to behave in ways that are less likely to lead to injury (physical and psychological capability). Conversely, pressures on productivity (physical opportunity) may reduce willingness of workers to take proper precautions (reflective motivation) thereby creating a community of unsafe practice that spreads through the processes of peer to peer role modelling (social opportunity) that means that new members of staff do not adequately acquire the competencies to do their job safely (psychological capability).

Deriving a behavioural diagnosis using the COM-B model has the potential to provide a comprehensive explanation of behaviours related to safety since it encompasses automatic processes as well as conscious deliberative processes, practical influences such as time and resources upon behaviour, as well as the complex web of social influences (e.g. what's normative), and the physical and psychological capabilities such as knowing why a behaviour is important and knowing how and possessing the skills to do it. It also specifically includes system level influences and emphasizes that behaviour is the product of, and therefore potentially influenced by, interventions at multiple levels of influence.

14.7 Intervention Design Using Intervention Functions

Once a behavioural diagnosis has been identified, the next stage is to design an intervention to influence the behaviour in the desired direction. The BCW describes nine categories of intervention classified by their function. These intervention functions and their definitions are described in Table 14.1. When designing intervention using the BCW the designer is encouraged to think about the entire range of possible ways to influence behaviour, not just the obvious ones or those with which they are most familiar as a result of professional training or experience. Nevertheless, it is clear that different intervention functions are more suited to influencing different forms of behavioural influence (e.g. training is an appropriate way to help people acquire the physical capability to perform a behaviour whereas persuasion will be ineffective). More information on the relationship between the COM-B domains and the interventions functions can be found in Michie et al. (2014).

Williams (2015) describes how the BCW was used to understand the positive impact of an environmental restructuring intervention on the incidence of road traffic accidents in a manufacturing company. The company fitted vehicles with driver-performance trackers as a means to increase productivity. Driver performance measures included 'fuel efficiency' and 'sympathetic braking/accelerating' and each driver's performance was visible to their co-workers on a live display screen in the transport office. The unintended impact of this was a dramatic reduction in the occurrence of road traffic incidents. Interviews with the drivers in the company revealed that this change to the work environment (an environmental restructure intervention) changed the norms by which drivers judged their driving (tackling social opportunity) as well as providing motivation to be the 'best driver' (boosting reflective motivation) as well as providing feedback to drivers about their performance which they used to improve their performance (improving capability).

Table 14.1 BCW intervention function definitions and examples [adapted with permission from Michie et al. (2014) and Williams (2015)]

Intervention function	Definition	Example of intervention function
Education	Increasing knowledge or understanding	Providing information on risks associated with non-compliance with machine operating instructions
Persuasion	Using communication to induce positive or negative feelings or stimulate action	Using images or stories drawn from real life accidents to induce the desire for compliance
Incentivisation	Creating an expectation of reward	Scheme to acquire benefits in return for compliance with behaviours related to safety
Coercion	Creating an expectation of punishment or cost	Loss of in-work benefits if found to be violating safety principles
Training	Imparting skills	Dynamic risk assessment skills
Restriction	Using rules to reduce the opportunity to engage in the target behaviour (or to increase the target behaviour by reducing the opportunity to engage in competing behaviours)	Prohibiting entry to certain areas of the plant.
Environmental restructuring	Changing the physical or social context	Changing work teams to provide social influences
Modelling	Providing an example for people to aspire to or imitate	Using shopfloor, peer coaches as part of manual handling training
Enablement	Increasing means/reducing barriers to increase capability (beyond education and training) or opportunity (beyond environmental restructuring)	Behavioural support for smoking cessation, medication for cognitive deficits, surgery to reduce obesity, prostheses to promote physical activity

14.8   Using Policy to Change Behaviour

Organisational context is a powerful determinant of safety culture and decisions made by those in power will have an important influence on the expression of safety-related behaviour. The BCW identifies seven distinct types of influence which can be leveraged by authorities to influence behaviour. These include:

- 'communication/marketing' (using print, electronic, telephonic or broadcast media);
- 'guidelines' (creating documents that recommend or mandate practice);
- 'fiscal' (using the tax system to reduce or increase the financial cost);
- 'regulation' (establishing rules or principles or behaviour and practice);
- 'legislation' (making or changing laws);
- 'environmental/social planning' (designing and/or controlling the social environment);

- and 'service provision' (delivering a service).

As with intervention functions, the relevance of each different policy function will vary according to the nature of the behaviour to be changed and the context in which it occurs. Nevertheless, the inclusion of a policy level of influence is an important differentiator for the BCW compared to other behaviour change frameworks, especially when applied to a sector where organisational culture has been shown to have such an important influence on the uptake and spread of initiatives to improve safe practices.

14.9 Using Behaviour Change Techniques Within Intervention Design

Intervention functions describe the different ways in which behaviour is influenced (e.g. through coercion and training) but they fall short of describing the specific techniques that are employed to bring about change. Behaviour Change Techniques (BCT's) are the observable, replicable, irreducible components of an intervention designed to change behaviour, for example, goal setting and self-monitoring. It is possible to specify the content of behaviour change interventions by listing these active ingredients using a hierarchical taxonomy of behaviour change techniques that has been scientifically developed through a rigorous process of expert consensus review (Michie et al., 2015). The Behaviour Change Technique Taxonomy V1 (BCCTV.1) is available online as a web and app-based resource as well being described in Michie et al. (2014). Specifying behaviour change interventions at the level of BCT's has the potential to allow researchers and practitioners to understand with more precision how different techniques may be related to outcomes. This may be useful for designing more cost-effective interventions since elements that are proven to be unrelated to outcomes can be discarded, or for identifying why similar interventions have different results in different contexts of delivery.

14.10 Potential Applications of the BCW Methodology for Industrial Safety

As a relatively new framework the utility and the effectiveness of the BCW is yet to be established in the field of industrial safety. Whilst it is for researchers and practitioners in this sector to decide for themselves the potential applications of this framework the following options may be useful starting points for further development based on its application to other areas:

- Using the BCW to design and develop curricula to embed safety considerations into the induction and training programmes for workforce development. For

example, the COM-B model could be used to audit or develop training programmes to ensure that workers are equipped with the three necessary conditions for working safely; having the skills to carry out task safely (capability);
the physical and social resources to do work safely (opportunity), and the sense
that working safely is a core part of what makes a 'good' worker (motivation).

- Adapting the BCW as a methodology for 'safety professionals.' The BCW
  provides a comprehensive behaviour change methodology that enables professionals tasked with safeguarding against risks to develop interventions in a
  systematic way, considering the entire range of possible influences on
  risk-related behaviour.
- Using the BCW as an organising framework for research into 'what works and
  for whom' in relation to interventions to improve industrial safety. The BCW
  and BCTTV.1 provide a methodology and tools to allow for greater precision in
  specifying the content of behaviour change interventions to improve safety in
  industrial contexts.

14.11 Conclusions

Like many other fields in applied behavioural science, the field of industrial safety
is in danger of being overwhelmed by the proliferation of theories and frameworks
that can be brought to bear on the perennial challenge of ensuring that workers
behave in ways that minimize the risks to self, others and the environment. The
emerging science of behaviour change, and specific frameworks such as the
Behaviour Change Wheel, create opportunities to integrate the valuable insights
from diverse disciplines using a single framework that has cross-disciplinary
appeal.

References

Abraham, C., & Michie, S. (2008). A taxonomy of behavior change techniques used in
    interventions. *Health Psychology, 27*(3), 379–387.
Fleming, M., Lardner, R., 2002. Strategies to promote safe behaviour as part of health and safety
    management systems, contract research report 430/2002 for the UK Health and Safety
    Executive.
Hollnagel, E., Wears, R., Braithwaite. (2015). From Safety-I to Safety-II: A White Paper. The
    Resilient Health Care Net: Published simultaneously by the University of Southern Denmark,
    University of Florida, USA, and Macquarie University, Australia.
Hopkins, A. (2006). What are we to make of safe behaviour programmes? *Safety Science, 44*(7),
    583–597.
Michie, S., Atkins, L., & West, R. (2014). *The Behaviour Change Wheel: A Guide to Designing
    Interventions*. Great Britain: Silverback Publishing.

Michie, S., & Johnston, M. (2011). Behavior change techniques. In M. Gellman & J. Turner (Eds.), *Encyclopedia of behavioral medicine*. New York: Springer.

Michie, S., Richardson, M., Johnston, M., Abraham, C., Francis, J., Hardeman, W., Wood, C. E. (2013). The behavior change technique taxonomy (v1) of 93 hierarchically clustered techniques: building an international consensus for the reporting of behavior change interventions. *Annals of behavioral medicine, 46*(1), 81–95.

Michie, S., van Stralen, M., & West, R. (2011). The behaviour change wheel: A new method for characterising and designing behaviour change interventions. *Implementation Science, 6*(1), 42.

Michie S, Wood CE, Johnston M, Abraham C, Francis JJ, Hardeman W. (2015). Behaviour change techniques: the development and evaluation of a taxonomic method for reporting and describing behaviour change interventions (a suite of five studies involving consensus methods, randomised controlled trials and analysis of qualitative data*). Health Technol Assess, 19*(99), 1–188.

Williams, C. (2015). Behavioural safety – intervening in a useful way. ttp://www.shponline.co.uk/behavioural-safety-intervening-in-a-useful-way/.

Chapter 15
Power and Love

Recognizing the power of 'those who make' to achieve enhanced (safety) performances, through dedicated spaces for debate

Nicolas Herchin

Abstract Building on other contributions, this chapter highlights how safety is a situated activity, which relies greatly on non-technical skills. As such, professionalizing in safety implies creating spaces for debate. In this context, the question of power is key to consider: indeed, professionalization is first a matter of identity, which in turn questions power, be it formal or informal. One of the key question is: 'how to cope with increasingly powerful specialists in support functions?'. As an attempt to answer it, this chapter argues that shifting from a 'love of power' to the 'power of love' is the key to liberated organizations in which (safety) performances are enhanced. Giving more power and consideration to working teams and middle managers in the field by creating space to discuss rules and practices is a first step to doing so. A second, more in-depth, step implies a change of paradigm from a 'simple' steering of safety indicators to a broad empowering of employees, giving them vision and autonomy to do their jobs. This involves a "liberation" process by which the classical vision of hierarchal structures is reversed, and the importance of learning and knowledge is acknowledged as a key source of motivation.

Keywords Identity · Empowerment · Spaces for debate

15.1 Introduction

In introduction, let us recall the conclusions and key take-aways of the seminar[1]:

---

[1]The two-day international workshop mentioned in the preface, organized by FonCSI in November 2015 and highlight of the project that led to this book (editors' note).

N. Herchin (✉)
ENGIE Research & Technologies Division, Paris-Saint Denis, France
e-mail: nicolas.herchin@engie.com

C. Bieder et al. (eds.), *Beyond Safety Training*, Safety Management,
https://doi.org/10.1007/978-3-319-65527-7_15

1. Safety is only **one of the aspects of doing things well**, one way to arbitrate. There are others.
2. One of the key issues regarding professionalization in safety lies in the difference between **'expert' knowledge** (i.e. coming from safety support functions) and **'field' knowledge**. Expert knowledge should be used in situations that are not known, whereas it is often used in situations where it is not needed.
3. Safety has to do with **identity**. Call it Community of Practice, Craft group, trade… The importance of professional groups producing safety rules which are not of the same type as rules written by experts should be emphasized. In this respect, there is a clear **need for physical and temporal spaces** to foster meetings and discussions from and by field operators, in order to build safety rules.
4. One cannot talk about professionalization in safety without talking of the **attributes of safety culture**: having good professionals implies that organizations have the capacity to listen to bad news, fostering debates and controversies. Safety culture requires humility, as well as recognition that no one has all the answers.

In this context, the **question of power** appears to be central in terms of capacity of operators—individuals and collectives—to fully embrace both the issues arising in their day-to-day activities, including—but not limited to—safety, and the solutions they can find to tackle them. Indeed, all the above-mentioned aspects of the question of professionalization are related to power issues: individual power (to act on one's reality in a given situation), formal and informal power (i.e. experts in support function vs. field operators), group power (deriving from a collective identity and a set of shared values), organizational power (i.e. power of the hierarchal structure in place).

This chapter aims to re-examine the academic contributions on the question of professionalization, trying to highlight the (sometimes hidden, but often omnipresent) role of power at stake in the process of professionalization. In a first part, the importance of creating spaces (both temporal and physical) for discussion and debate between operators is emphasized, in reference to Gherardi's, Flin's and Boccara's contributions. The second part builds on Hayes' and Ughetto's chapters to highlight how the notion of power acts as a central driver as far as professionalization is concerned. Finally, a synthesis of key findings and possible ways forward is attempted in the third part.

15.2 Professionalizing in Safety Implies Creating Spaces for Debate

15.2.1 Safety Is a Situated Activity…

As shown by Silvia Gherardi in her chapter on safety as an emergent competence, safety is a social competence which is realized in practice, socially constructed, innovated and transmitted. She proposes looking at "safety as a collective knowl-edgeable doing, i.e. a competence embedded in working practices." In other words, safety is a **situated activity**. Therefore, the safety culture of an organization or group of people is one of the distinctive features of professionalism, which itself derives from the way contextualised situations are apprehended by workers.

In this context, Gherardi sees safety as having the following characteristics:

- Safety and practice cannot be separated, as the former derives from the latter.
- "Safety is performed in, by and through social relations", including at the heart of it, language as an essential medium.
- Safety is rooted in "practical knowledge", i.e. based on tacit "competence-to-act" (vs. knowledge to act), emerging from individual and collective identity.
- Safety is "dynamic", "emergent from actions, and in constant evolution". "It is continually re-produced and negotiated."

This approach has two main consequences when considering safety:

1. One should consider work practices first, rather than separating safety out from local situations and contexts. This means privileging bottom-up, 'describing', approaches rather than top-down, 'prescribed', ones.
2. Focus should be made on the capacity of individuals to discuss situations and ways to tackle work, as safety is embedded in practices and emerges as a social construct.

In other words, when talking about safety, one has necessarily to acknowledge the importance of Non-Technical Skills (NTS), as introduced by Rhona Flin in her chapter. NTS are indeed are the core of field expertise regarding work tasks, and working safely (and efficiently).

15.2.2   ...Which Relies Greatly on Non-technical Skills

For Flin, **Non-Technical Skills (NTS)** are a necessary complement to expert, technical skills in order to enhance safety and efficiency. In her mind, and in the continuity of Gherardi's thoughts, focus should be made on workplace behaviours and work as performed on a given task. Indeed, there is no other way out of the "safety bubble", that is safety being seen, taught and performed as isolated and distinct from normal operations.

In short, taking NTS into account depends on the ability to access detailed data on the reality of what goes on in the field, with all its complexity. This may be achieved by involving human and social sciences experts to identify NTS, and performing training courses based on NTS, such as CRM (Crew Resource Management) approaches, which are task-related.

In the end, safety is only one aspect of doing things well, one way to arbitrate. There are others, all emerging from particular situations and contexts in which individuals and teams are bathed, mobilizing technical as well as non-technical skills to tackle situations and issues to the best of their ability. As a consequence, when considering safety as being embedded in practices, it becomes necessary to have a grasp of the 'global picture', i.e. to understand the entire panel of constraints leading to the arbitrations made.

This, again, links back to the importance of having identified spaces to allow discussion on practices.

15.2.3 As such, Professionalization in Safety Requires Space for Debate

In the light of the above, if organizations are to have good professionals, this implies that they must be capable of listening to bad news, fostering debates and controversies. Safety culture requires humility, as well as recognition that no one has all the answers. To put it in a nutshell, providing **physical and temporal space for formal and informal discussion** is fundamental to enhancing safety—and overall performances—in the organization.

As Vincent Boccara puts it in his chapter on safety training, most often organizations tend to focus training on theoretical situations; the problem of this approach becomes: how to "make people able to deal with real-world situations rather than only know and applicate exogenous standards." Therefore, safety is primarily a matter of organizing discussions and controversies inside the organization on how things (really) occur on the shopfloor.

Pascal Ughetto, in his chapter on empowering line managers, also emphasizes the importance of allowing room for controversies, in the way these are central to "exchange about the real activity, its risks, its opportunities, and therefore [to] accommodat[e] to actual working cultures." There is a need at every level for room to intervene. Every group has a role to play to construct the problem so it can be recognized, and result in "a continuous process of rule creation."

In other words, there is a clear need for fostering debates inside communities of practices, but also at the borders of these communities, on practices and 'the way to work around here'. Indeed:

- Encouraging (formal and informal) discussions between operators—i.e. allowing time and installing rituals in the organization to do so—consists in a bottom-up approach essential to embed safety in real-work situations.
- Language is the first, essential social medium through which expertise can be shared, and therefore built. For example, sharing NTS by mentoring activities, or by "storytelling" (Hayes) how unusual events in the past were managed differently.
- Discussing practices is key to gaining efficiency, be it on safety or any other aspect; questioning rules and procedures, in other words the prescribing system,

in the light of daily practices and particularly their variability, can prove highly efficient for the organization and is worth the time investment.

- Finally, trying to embrace the global picture also means questioning practices at the frontiers of communities, through interactions between the various departments in the organization, particularly support functions. A better balance is required between top-down, prescribed rules and bottom-up, built-up responses to various situations.

15.3 The Question of Power Is Important to Consider in this Context

15.3.1 Professionalization Is (also) a Matter of (Group) Identity

As Jan Hayes describes in her chapter, professionalization is more than simply a matter of training; it is first **a matter of identity**. Be it communities of practice, craft groups or trades…, we have seen the importance of having professional groups sharing practices and producing safety rules which are not of the same type as rules written by experts. These groups share, and build together a common identity, strongly based on the type of activities performed: 'what we do' is 'who we are'.

In this respect, as illustrated by Hayes, storytelling is a key component of social learning, as discussions on past events are full of learning material for peers. **Trust** is therefore of the greatest importance in order to create a favourable climate to make these discussions possible. This, again, emphasizes the need for spaces for debate.

In short, many employees in industries have a (shared) professional identity, emerging from social interactions. This identity remains largely unrecognized by organizations; yet, professionals ensure **mutual recognition** through various informal processes, all leading to professionalization processes. This identity, in turn, questions the role of power in the relations between different organization's entities.

15.3.2 Identity Questions Power (Formal or Informal)

For Jan Hayes, top-down, hierarchal approaches are prevailing in most companies, whereby managers set the "rules" and people at the bottom simply follow the instructions. Yet, professional groups have a certain **power**, which derives from their **identity**; this power is often informal and not fully recognized, but this question is key, as it can explain the relative inaction inside companies when it comes to improving safety, or any other performance. Indeed, as long as a decision does not fundamentally bring into question the identity (thus power) of a group of people (i.e. managers, safety experts, or operators), compromises can be found, although often leading to minute changes. But as soon as the identity and power of such a group is threatened, decision-making becomes more difficult, resulting in the preservation of a status-quo, thus a form of 'social peace' in the organization.

In this context, power translates firstly into *individual power*, i.e. the capacity of an individual to act in a given situation, towards a given goal, according to his or her values and perceptions, mobilizing technical and non-technical skills acquired from past experience, and taking into account various constraints, from which working in safety can be one of them. Secondly, *group power* derives from the collective identity of a community to which an individual belongs. Shared values and practices, storytelling and trust form the basis of this identity, which in turn translates into a shifting balance of power in the interactions between entities. Finally, these two forms of power can be *formal or informal* depending on whether their legitimacy is given by the structure (via hierarchal recognition e.g.) or by peers (recognizing expertise).

In this respect, the process of recognising professionalism (and thus power) is key. However, as Hayes pinpoints, research has shown the extent to which organizations rely on professional behaviour for ongoing safe operations and yet largely fail to understand or acknowledge this. Of course, power issues are at stake.

In brief, **competition for power** is emerging between entities in the organization, each seeking its own good according to its identity (Ughetto). But what about the common good, especially as far as safety is concerned? The common good of the company (e.g. in terms of cost of accidents), its members (e.g. victim of accidents), partners (subcontractors, clients, …), but also the common good of external stakeholders (third parties). In other words, is power from the field operators given enough place, and does it translate into legitimate authority when it comes to matters of safety? Or is power perceived as a danger by those who hold the formal power, such as support functions or managers?

15.3.3 How to Cope with Increasingly Powerful Specialists in Support Functions?

Jan Hayes recognizes the existence of an injunction inside companies towards operators to be "good professionals"; yet, this injunction is rarely accompanied by:

1. a will to better understand the levers that can be triggered to allow this, and
2. means of doing so for operators (i.e. time and space for discussion and "social learning practices", to begin with).

In short, "the forms of power in organisations increasingly limit the recognition of this expertise [i.e. of line managers as a support to real work activities and daily arbitrations]", as Ughetto states,

> In power relations within large organizations today, the power exercised by support functions – through the primacy of standards – deprives field managers of a great deal of *[leverage]* for action. This power, despite the diversion via participatory management, leaves little room for regular discussion of the relevance of organizational rules. However, rules – in particular safety rules – are not purely and simply "implemented": they need to be discussed.

Indeed, for central management, recognizing the power of professionals in the field means potentially putting in danger their own. From a manager's point of view, this can mean losing control to some extent. In their formal, "top-down", conception of work, employees should merely apply rules; consequently, allowing time for discussion is a pure loss of time and money.

In the case of support functions, giving power to local teams can mean endangering their position in the organization, potentially leading, once again, to a loss of authority or control. Indeed, safety experts, by doing so, take the risk, as they see it, of ultimately losing the very meaning of their job. This can prove a paradox in some sense, as safety experts may eventually prevent safety enhancement by seeking to keep control of safety-related practices. This introduces a very fundamental question: how can their roles be rethought in such a way to prevent them from feeling threatened?

To conclude, let us retake Ughetto's words:

> To achieve this [common good across the organisation], it needs to be accepted that (...) introducing organisation into day-to-day operations, and notably the organisation of safety, is **not about implementing organisational rules** and letting them operate unchanged for a given time; **the issue is organising, a continual activity of organisation.**[2]

This can be achieved by redistributing power within the organization and reinventing work habits, shifting from the perception of safety-as-a-constraint to safety as 'a way of doing things right'.

15.4 Shifting From' Love of Power' to 'Power of Love': The Key to Liberated Organizations in Which (Safety) Performances Are Enhanced?

To go a step further, let us introduce some possible suggestions for better managing this 'continuous activity of organisation' as phrased by Ughetto. To sum up, power issues at stake in every company based on hierarchal structures tend to limit the ability of operators to tackle situations and problems on their own. One of the first

---

[2]I emphasize.

steps is to give them more power to do so by creating spaces for discussion, as already stated in this chapter. Thus, the process of organizing work becomes more balanced between top-down, prescribed approaches, and a bottom-up vision of the real issues and best ways to tackle them.

Building on this first step, a change of paradigm may be introduced, by which power is shifted back to 'those who really make safety' (and more generally performance), considering their personal development and well-being as central. This leads to more autonomy in their daily decision-making, and in turn, to optimized professionalization processes, which are key to performance.

15.4.1 Giving More Power and Consideration to Working Teams and Middle Managers in the Field by Creating Spaces to Discuss Rules and Practices

Jan Hayes insists on the need to allow time for discussion of professional activities, as "professional learning is a profoundly social activity". Yet, "many organizations are reluctant to allow time for such activities." This is of course a question of time, but also of recognition that fostering safety goes through training good professionals, which in turn implies giving more power and consideration to local teams in the field.

As seen in this chapter, and in many other contributions, creating dedicated spaces is essential to allow discussion of rules and practices in the light of situations and constraints which are encountered by 'those who make' safety. This is firstly a matter of recognizing the importance of both the variability of situations met, and the daily arbitrations made by operators in order to maintain a good level of performance, safety included. But it is also a way to bring to light different group strategies leading to more or less safe practices, then putting into discussion these practices (Le Coze et al. 2012).

However, spaces for discussion are not enough; there is a clear need, in parallel, to support middle managers in coping with issues identified through feedback. As Ughetto rightly concludes, more power should also be given to middle managers

> to do something with the complaints of their teams, to analyse the work, its constraints, how the teams go about getting things done, and to make proposals to their line managers and their teams.

To conclude, still using Ughetto's words,

> the key question is therefore how much space today's organisations allow for experiment, for variability, and how much space they give middle managers to construct organisational rules, first of all by holding discussions within their teams and between those teams and support departments.

15.4.2   Towards a Change of Paradigm: From Steering Safety Indicators to Empowering Employees, Thus Giving Them Vision and Autonomy to Take on Their Jobs

Undoubtedly, there is a need for more debates on work. But even more so, the **work model** at stake in the company can be questioned: what should be improved in the system? Indicators, or practices? Thus, the question of shifting power back to the operators, leading to a paradigm change.

15.4.2.1   The Paradigm Change: Reversing the Classical Vision of Hierarchal Structures

As introduced earlier in this chapter, top-down, hierarchal, approaches most often prevail, whereby bottom-line employees follow the rules dictated by managers. This often leads to performance destruction, as the motivation expectations of the employees are not met: indeed, if operators are told what to do and how to do it, where is the motivation to work, and learn from its work? Of course, some space for self-organization can be found in today's companies; yet, the trend is that 'instructions' and 'rules' come from the 'top', representing ever more constraints hindering motivation to learn and work.

Given this widely prevalent hierarchal model, the paradigm change we propose comes from reversing this work structure by shifting back the power to field operators, as theorised by Brian Carney and Isaac Getz in their book "Freedom Inc.", based on many concrete examples of "liberated" firms. In such 'liberated companies', much more autonomy is given to employees in their daily work. In a climate of trust, employees are given the capacity for self-direction and self-motivation, or in other words the means and power to act and decide without referring to management or transverse functions. The latter then see their roles change, from a 'command and control' role to a more 'humble servant' role, in which they bring support and facilitation to 'those who make'.

In other words,

a place of work focusing on respect and liberty is much more natural than an environment based on mistrust and control. (...). Every morning, employees go to work, but many of them prefer saying they are going to take pleasure following a common dream, putting in place their initiatives. And coincidentally – or perhaps naturally – these organizations realize continuously better performances as their competitors. In other words, respect and liberty are keys to pleasure and success (Carney and Getz 2016).[3]

---

[3]"Freedom Inc.", preface of the new 2016 edition (author's translation).

15.4.2.2 The Importance of Learning and Knowledge, as a Key Source of Motivation

In short, self-motivation and self-direction are the key to empowering employees towards improved performances. And one of the key sources for motivation is learning; performances thus come in a great part from the capacity to learn, and share knowledge.

In one of his articles on the 'knowledge economy', Idriss Aberkane emphasizes the three rules governing knowledge exchanges:

1. Knowledge exchanges are positive sums: sharing knowledge means multiplying it.
2. Knowledge exchanges are not instantaneous; they take time.
3. Grouping knowledge creates knowledge: knowing A and B together is more than knowing A and knowing B separately.

On that basis, the flow of knowledge grows in proportion as a product of attention and time. And this product basically translates into love. Indeed,

> in what circumstances do we give all our attention and our time to someone? When we are in love! We never learn as rapidly as when we are in love with some piece of knowledge (…). We should seriously take this into account in our education and in our companies (Aberkane 2014).[4]

In other words, when talking of professionalization in safety, one should bear in mind the importance of motivation, which itself mostly comes from a love of learning and knowledge. In the end, liberated companies in which employees are self-directed are most surely the best environments to promote love, therefore spaces of dedicated time and attention to allow the knowledge of situations and issues flow, in all their complexity and variability…

As a conclusion, balancing power inside companies through their 'liberation' should be a key move to promote love, thus shifting from the 'love of power' of some individuals to 'power of love' for all. In our opinion, enhancing knowledge and performances, including safety, clearly comes at that price.

15.5 Conclusion

This chapter has attempted to synthesize various contributions, highlighting several key elements:

- The need for contextualised, bottom-up approaches, as safety is a situated activity where non-technical skills play an essential role (Gherardi, Flin).

---

[4]"L'économie de la connaissance est notre nouvelle renaissance", article from Idriss Aberkane (author's translation).

- The importance of giving the power back to field operators by promoting discussions between peers about practices, stories, rules, constraints, etc. (Boccara, Hayes, Ughetto).
- The key role of middle managers as a relay for processing feedback from the field (Ughetto).

It also appears that the question of power is key in the process of professionalizing employees in safety. Indeed, on the one hand it appears that physical and temporal spaces to allow practices to be discussed are essential for fostering improvements in a "continuous activity of organization" (Ughetto). Yet, it seems on the other hand that in most companies today the power of employees to do so is undermined by the weight of hierarchal structures tending to hinder initiatives. In other words, there are things which cannot be discussed, especially about the way to work safely, which appears to be all the contrary of professionalism.

In this context, and as a conclusion, we suggest two paths for improvement:

1. In a first one, we suggest, on a continuous improvement basis, to give more power and consideration to working teams and middle managers in the field by creating (physical and temporal) spaces to discuss rules and practices. This links back to Ughetto's proposal to shift the way power is organized or distributed, allowing time for discussion and sensitizing support functions and middle management to the importance of listening to feedback and do something about it.
2. In a second one, we propose a paradigm change, questioning the very work model of most companies. Building on Carney's and Getz's examples, we believe that following a process of "liberation" in companies would facilitate enhanced knowledge sharing and collective learning. This in turn would enable the shift from a 'love of power' to the 'power of love', leading to better performances, including safety performances.

Whichever way is chosen in the end, acknowledging the power of 'those who make' remains the key for enhanced (safety) performances.

15.6   Disclaimer

The views and opinions expressed in this chapter are the sole responsibility of the author and may not reflect those of ENGIE.

References

Aberkane, I. J. (2014, June 4). L'économie de la connaissance est notre nouvelle renaissance. *Huffington Post.* (http://www.huffingtonpost.fr/idriss-j-aberkane/economie-de-la-connaissance_b_5443212.html. last search 19/04/2016).

Carney, B. & Getz, I. (2016). *"Freedom Inc."*—French version *"Liberté et Compagnie"*. Flammarion.

Le Coze, J-C., Perinet, R. Herchin, N. & Louys, P. (2012). *To describe or to prescribe? Or both?* 6th International Conference Working on Safety "Towards safety through advanced solutions", 11/09/2012–14/09/2012, Sopot, Poland.

Chapter 16
Beyond Safety Training, Toward Professional Development

Synthesis and food for thought

Caroline Kamaté, Hervé Laroche and François Daniellou

Abstract Professional development in safety lies at the crossroads of various logics, each with their own objectives, limits and power games. The arbitration and choices that are made at different levels (individual, collective and organizational) are therefore subject to constraints. It is of major importance to be aware of these constraints, to take them into consideration and recognize them in order to identify the levers for improvement in safety performance. This chapter synthesises the main findings from the book, highlighting what is currently considered to be at stake in terms of safety training, in the industrial world (industry and other stakeholders such as regulatory authorities), and offers avenues for further research.

Keywords Continuous learning · Vocational training · Situational simulation

16.1 Introduction

Despite the increasing attention given to safety training, safety results—notably in industrial sectors where they are already well-developed—seem to have reached a plateau in companies in charge of high-risk activities. Why are accidents still occurring, despite the significant improvements observed? What about the return on investment? Should we provide more safety training? Should we train people differently? These were the underlying issues of the concerns expressed by FonCSI's industrial partners. In light of this harsh assessment, researchers from both the industrial sector and the academic field engaged in an 18-month discussion coupled with a 2-day international seminar. The aim was to explore new avenues to improve industrial safety in companies and give it a more 'professional' dimension.

C. Kamaté (✉) · F. Daniellou
FonCSI, Toulouse, France
e-mail: caroline.kamate@foncsi.org

H. Laroche
ESCP-Europe, Paris, France

© The Author(s) 2018
C. Bieder et al. (eds.), *Beyond Safety Training*, Safety Management,
https://doi.org/10.1007/978-3-319-65527-7_16

The resulting book stands out not only for the diversity of its contributions but because it reflects the debates that have been engaged between the authors. It offers a critical analysis of safety training as defined, envisaged, and organized in at-risk industrial sectors. The challenge is that professional development in safety lies at the crossroads of various logics, each with their own objectives, limits and power games. The arbitration and choices that are made at different levels (individual, collective and organizational) are therefore subject to constraints. This book highlights how important it is to be aware of these constraints, to take them into consideration and recognize them in order to identify the levers for improvement in safety performance. The project that led to this book enabled a number of links between different disciplines, different industries and countries to emerge, and clearly identified points of convergence between the various contributors. Main findings and subsequent stakes and levers that have been identified for improvement are summarized below. At the end, we propose a research agenda aimed at opening new avenues for reflection and possible field experiments.

16.2   Safety as a Dimension of Professional Development

First, the authors agreed on the fact that safety should not be addressed as an isolated dimension. Safety is a feature of everyday working practices,[1] from normal to crisis situations. They found consensus on points that initially appeared quite extreme, such as the impossibility of separating safety know-how from professional know-how (safety skills from professional skills). *Safety is one 'result'—among others—of 'doing things right'.* This clear assumption that safety is an integral aspect of professionalism raises the issue of the general perception the organization has of the link between safety and professional development.

16.2.1   The 'Good Professional'

A good professional would be better equipped to make the most appropriate choices in any situation—one which might impact safety as well as other performances—taking into consideration various constraints. But what is a 'good professional'? The criteria differ depending on who is determining it. Although from the peers' viewpoint being a good professional has something to do with the identity of the trade, the identity of the work collective, from the viewpoint of the organization, professionalism is defined in a much more top-down manner. As an example, the

---

[1]It can be continuously produced/ threatened.

competency framework is defined by human resource managers for human resource managers who use it as a job management tool rather than as an activity management tool. The term 'profession', according to the sociology of work and professions (and to its English meaning), implies undergoing training recognized at the State level, in a sector that has regulated access, with the possibility of a life-long career and finally the existence of a body of 'professionals'. In some companies tradespeople claim to be professionals, although this is contradicted by the organizational structure, which does not meet the above conditions, notably in terms of career opportunities within the so-called profession. The 'management bubble' has developed in isolation from daily practices. What is the relationship between these two forms of identification and assessment? How can we help to reconcile the job as conceived by human resource managers and the job as actually done? The main issue is the connection between the viewpoint from the top and the one from the bottom.

16.2.2   Time Issues

There is much movements within trades. And there is a contradiction between this rapid turnover and the time needed to make a 'good professional'. In a rapidly-changing environment, companies look for the minimum skills, which goes against the idea of trade as an art. Should the turnover be slowed down? Should the adjustment of industry be promoted?

16.2.3   Safety Training for External Justification

Mostly, safety training courses are focused on rules, procedures, fuelled by experts' knowledge and standards and taught in a way that is disconnected from the professional gestures. They are usually designed to respond to high external justification issues, 'external' being here understood to have several meanings: supervisory authorities, media, public opinion, which are somehow reflected by internal support functions. Negotiation issues should not be neglected: training can also be a pacification tool towards unions. This justification system involves mainly specific Health & Safety actions to 'tick the boxes' that are imposed by external prerequisites. Compliance to standards is mandatory and cannot be avoided. However, there is a decoupling between the standardized, certified stratum and the stratum of workplace routines. Strengthening the internal normalizing policies exacerbates this decoupling. We would tend to suggest that companies limit their investment in certified trainings to those that are required by law.

16.3 Pedagogical Precautions

Working on the assumption that it is understood that safety emphasizes professionalism, promoting its incorporation into training programmes requires pedagogical precautions.

16.3.1 Safety and Real-Life Working Situations

Even contributing scholars whose research fields are not identified by theories of activity expressed a strong interest in actual working situations (situated action). Thus, it is necessary to recognize and explain constraints in order to understand tradeoffs that are made in daily practices, either in nominal or in degraded situations. This underlines the importance of introducing safety as a component of vocational training, which is centered on the technical gestures. But depending on the working situation and its induced constraints, different kinds of (safety) knowledge will be mobilized. Frontline managers are often caught between the knowledge of experts and tradespeople, between safety based on rules and managed safety, with little leeway. Some situations that have never before been encountered, so called unexpected situations, would require experts' knowledge in order to be resolved. On the contrary, in many cases where the situation can be anticipated, reference to expert knowledge will be imposed although it would not be necessary. What are the possible spaces of articulation between these different 'poles' (experts/trades)? The use of simulation to prepare trainees not only to normal but also to degraded situations favors both consideration of rule-based safety and the development of a certain ability to manage safety. Then the issue of transfer is of major importance: what happened during the training session? What will actually be implemented in real work situations?

16.3.2 Professional Development as a Whole, not Limited to Training Sequences

Training in safety is about promoting the development of 'good professionals' at large. 'Technical' training courses that include safety aspects and focus primarily on improving the performance of practices are the most effective ones for anchoring 'good' behavior in professional practice. It is therefore of importance to focus not only on the training sequence. The learning process must be considered in its entirety, as a continuum covering various places, including critical moments, in a more or less enabling environment: psychological support, a recognized right to make mistakes, room for debriefing, debating, and for reflective practice. The general context of the working situation must be taken into account. The logic of a professional 'journey' should be adopted: reception of newcomers, attention given to the narratives of elders, support and companionship.

16.4 Beyond Training Issues, Organizational Stakes

16.4.1 Give More Room to the Professional Figure

The contents of the book confirm that the companionship and example conveyed by field managers are ingredients that largely contribute to actual professional development. However, these dimensions are often poorly recognized and given little accompaniment by the organization. The latter has to recognize that the definition of safety is also built through exchanges among peers. However, spaces that would allow collaboration, discussion about practices, where contradictions encountered in real situations could be explained and debated, where compromises could be made—at least partly—explicit, are in worryingly short supply. The creation or promotion of such visible and known spaces, a sort of recognized 'parentheses' should be considered. This would be useful to promote settings between a standardized practice of safety and a professional safety-appropriate practice. Frontline managers should have enough flexibility and receive the support of their hierarchy to implement such spaces.

16.4.2 But Avoid the Seductive Trap of the 'Professional Hero'

The trend to go back to actual work, to relocate the 'good professional' at the core of the skills topic, is a result of the pendulum swinging back after oscillating far towards the prescriptive side. But there is a risk in the glorification of the professional, in having the feeling that the worker's perspective is necessarily the truth. It could mean that mechanisms are missed, generating collective blindness. It is not desirable to value a figure of 'professional heroes'. Collective reflection on working practices and the framing by the group of individual initiatives, are an essential issue of learning.

Another approach is to set a target where safety is part of the rules of the trade. But it is not so simple. Agents can reject the safety injunction if the standards do not reflect what they consider to be 'real safety'—which does not mean they do not care about safety. Some workgroups have developed defensive strategies that lead to risk-taking. The defense of the trade can be in contradiction with standards from elsewhere, hence the difficulty to 'let safety in'. This requires professional training in which agents will 'rework' safety from the viewpoint of the trade: to be a 'good professional' is to get the trains running on time AND safely.

16.4.3 Reinforce Collaboration

Professional development requires the reinforcement of transverse collaboration skills, which implies knowing enough about the work of others. However, knowing about the jobs of others definitely does not necessarily lead to harmonious relations: depending on the organizational context, it can also be used to better 'trap' others. Designing training schemes together should help to establish some trust between professionals and organizations on matters related to safety.

16.5 Towards a Research Agenda

At the outset of this project, it was obvious for us that much research is still needed. Developing an exhaustive research agenda is beyond the scope of this chapter. However, we wanted to sketch out a short list of themes that, in our view, are worthy of attention from researchers (and practitioners!) and could greatly benefit from empirical research. In short, in this book, we have developed a novel way of approaching the issue of safety training and safety professionalization. However, for the purpose of demonstration, we did this at the cost of some simplifications and deliberately left to one side some important factors and actors. The agenda described above is mostly about broadening the picture and building a more real- istic, though also more complex, view of the issue.

16.5.1 Top Managers and (Safety) Professionalism

As often underlined in this book, our approach to professionalization questions the standardized representations of operators. Professionalism is often a claim made by members of the 'operational core' of the company as a defense, a protest against what is perceived as an excessive top-down or bureaucratic control. Paradoxically, as we have shown, top management and human resource executives often also support and promote an official discourse that calls for more 'professionalism'. Obviously, derivatives of the word 'professional'—professional, professionalism, professionalization—attract various actors with various, if not conflicting, world- views and purposes. There is little chance of consensus developing around these words and their implications. However, an overall picture of the way they are used, by whom, with which underlying meanings, in which purposes, could help us understand the 'system of professionalization' within organizations. The techno- logical developments (or rather, the beliefs and expectations related to them) and managerial philosophies ('the future of management') should be incorporated into this analysis, along with speculation about the 'future of work'. Developing such a picture would help find answers to a tricky question: what is expected—or feared—

from all this 'professional /ism /ization'? On what grounds can we design or arrange fruitful organizational dynamics, undoubtedly not without conflict, but with an expected positive return for most?

16.5.2 Evaluating the Efficiency of Standard Methods and Practices for Safety Training

One of the starting points of this project, based on recurring complaints from industry experts and managers, is that standard practices of safety training are of marginal assistance in industries and companies with a developed safety culture. Although the quality of the training is evaluated, its objectives have yet to be thoroughly assessed. More specifically, to what extent are these practices efficient? The potential adverse effects of indicators must be kept in mind; any prescriptive approach must be contextually reinterpreted and adapted to give sense to professional development. Evaluation implies looking at the whole learning process. It is important to go up the chain, to open the black box to see what actually worked, which is not necessarily the *training* per se. What exactly does 'a developed safety culture' mean regarding this issue? When a company has reached this 'developed' stage, should standard practices be reduced to the minimum level of mandatory requirements? When a company has not yet reached this 'developed' stage, should approaches be reinforced or complemented by additional, innovative methods and practices? While we felt that in this project priority should be given to industries and companies that already have achieved a high level of safety and are already equipped with a large array of safety training programmes, we are in need of a clearer picture of the benefits of safety training in low or average stages of development.

16.5.3 Rejuvenating Standard Safety Training

Given that standard methods of training are here to stay because of regulatory or accountability requirements, it can be argued, taking another angle, that some efforts should be devoted to getting the best out of them. In other words, what can be done to make them more efficient, within the framework of standard practices? For instance, what can be done to prevent routinization and ritualization? How could standard safety training get away from the excessive standardization of safety training and maintain contextual relevance? Which micro-practices, local innovative methods, additional or alternative resources could be introduced, with a low additional cost and within existing occupational and organizational constraints?

16.5.4 Reconsidering the Contribution of Safety Professionals

Another openly assumed bias of our approach was to take safety professionals (Health and Safety departments) out of the picture and deliberately focus on sharp end workers. Obviously, this is an oversimplification of the issue of professionalization. From an organizational viewpoint, safety professionals interact with the question of operators' skills in a complex manner. For one thing, fostering the professionalization of operators in the sense that we advocated in this book could be seen as a threat to the expertise and power of safety professionals. Conversely, if, as we believe, safety professionals face frequent managerial inertia or reluctance within the course of their mission, they might be truly interested in gaining support from operators and first-line managers, and a revised approach to safety professionalization could be an opportunity in their own eyes. Are safety professionals members of a techno-structure that produces standards and control devices, or are they brokers of ideas and practices between managers and operators? Most likely, a wide variety of situations may coexist, depending on the industry, the company, the occupations themselves, and the spirit of the time. With regards to this organizational view, safety professionals are likely to influence professionalization from inside through the kind of safety principles, philosophies, tools, etc., that they favor because of their own training and expertise. Also, they may influence professionalization from the outside, as members of professional societies and networks, having a direct relationship with regulators and producers of norms. A better assessment of the roles of safety professionals is needed.

16.5.5 Putting Other Actors Back in the Game

For the sake of simplicity, we also chose not to investigate the role of various actors that have a stake in the issue: unions, human resource departments, managers, regulators, etc. Safety training and professionalization are part of a wider system and national contexts have to be taken into account. In the case of France, for instance, occupational training in general is one of the principle battlegrounds between unions, governmental bodies and state agencies, and human resource departments. To give an example, irrespective of its nature and efficiency, safety training, expressed in terms of volume and budget, is used by companies as a demonstration to regulatory bodies and unions of their commitment and compliance. A different approach to safety professionalization, as advocated in this book, would have to fit into this wider system—or somehow find ways to escape from its struggles. Additional research is needed to explore the general dynamics of safety professionalization, according to the industries and national contexts, and to decipher which alternative views could be promoted and implemented within, alongside, or against such dynamics.

16.6   To Conclude

Since they affect both training and the participatory dimension of the corporate/unit culture, the inflection points suggested by these findings represent high stakes for the organization. It therefore cannot be considered desirable to advocate for a rapid generalization. Instead we would encourage the multiplication of experimentation, whether at the site level or with regards to one particular trade, depending on the company's current challenges. We suggest that these experimental approaches be implemented in a negotiated framework, and placed under observation with scientific support, with the aim of capitalizing and transferring the results.

Printed in the United States
By Bookmasters